PERSPECTIVES ON INTELLIGENT TRANSPORTATION SYSTEMS (ITS)

PERSPECTIVES ON INTELLIGENT TRANSPORTATION SYSTEMS (ITS)

Joseph M. Sussman
Massachusetts Institute of Technology
Cambridge, Massachusetts

 Springer

Library of Congress Cataloging-in-Publication Data

Sussman, Joseph M., Ph. D.
    Perspectives on intelligent transportation systems/Joseph M. Sussman.
        p.   cm.
    Includes bibliographical references and index.
    ISBN 0-387-23257-5 (alk. paper)
        1. Intelligent Vehicle Highway Systems.   2. Highway communications.   3. Electronic
traffic controls.   I. Title.

TE228.3.S87 2005
388.3'12—dc22

                                                                    2004062643

A C.I.P. Catalogue record for this book is available from the Library of Congress.

ISBN: 0-387-23257-5

Printed in the United States of America

9   8   7   6   5   4   3   2   1

springeronline.com

To my family—all the generations

Preface

"Individuals in every generation, at least since the birth of the Industrial
Revolution, have thought they were watching the most rapid changes ever
seen in history. They have all been correct."

—Robert Bruegmann
Preparing for the Urban Future

Since 1988 I have had the good fortune of being heavily involved in the
Intelligent Transportation Systems (ITS) program in the United States and abroad.
I began as a founding member of Mobility 2000, the predecessor organization to
ITS America. In 1991–92, while on sabbatical from MIT, I had the exceptional
experience of participating in the definition of ITS through the development of the
Strategic Plan at ITS America, entitled "A Strategic Plan for IVHS in the United
States". I served thereafter on the Coordinating Council of ITS America, and from
1995 to 2001 as a member of the Board of Directors of that organization. My
service on the ITS Committee of the Transportation Research Board has been
important to my ITS maturation. So through those activities, research and teaching
at MIT, and various consulting assignments, ITS has been an important part of my
professional life for quite a while.

Through that involvement, my view of ITS has evolved. Initially my perspective was that ITS was of crucial importance for the surface transportation field,
providing as it does, capacity through advanced technology rather than additional
conventional infrastructure. But over the years I have come to view ITS more

broadly. It can be a lens through which to view other important issues related to the transportation field and even the world beyond. This book takes both perspectives. Section I introduces ITS through an article of mine, "ITS: A Short History and a Perspective on the Future". This represents my early view: ITS as a surface transportation efficiency and effectiveness program. However, near the end of that article, we begin to consider the strategic changes ITS has created.

Sections II and III then consider the broader view. In Section II we consider ITS as a motivator and enabler of organizational change. This includes evolving relationships between various levels of government, as part of our expanded definition of infrastructure—the *transportation/information infrastructure*—and as a mechanism to develop effective *regional* transportation systems. In Section III we view ITS in its role as a change agent for the *transportation profession* and for *transportation education*.

Section IV takes a different tack. For five years I wrote a column for the *ITS Quarterly*, each about 800 words in length. They were intended to capture the essence of a timely ITS-related issue. By looking at those columns in their entirety, perhaps the reader can get a sense of evolving questions in the field, at least as seen by one observer. These columns cover questions such as how one teaches ITS, safety implications, the role of ITS in the 21st century, and ITS and megacities in developing countries. They represent quick probes into ITS, each at a particular point in time. Most, if not all, of the issues persist to the current day.

Section V brings us to the present and looks into the future. In 2000 I worked with the Joint Program Office (JPO) of the United States Department of Transportation (U.S. DOT) on a volume entitled "What Have We Learned About Intelligent Transportation Systems?" I was charged with synthesizing the findings of seven groups of researchers assigned to describe the state-of-the-art in ITS deployment in various areas. Section V captures my contribution to that effort, an overall summary of the work of my co-workers on this project, and some overarching final conclusions about what we face in the future in the ITS field. Section V also contains an article of mine dealing with important transitions in the transportation field, many enabled or even required by ITS, and a retrospective article on the 1991/92 ITS Strategic Plan considering what we have learned since that time.

So what this book tries to do is present the fundamentals of ITS, *and* consider how ITS has changed the world of transportation beyond the transportation effectiveness and efficiency changes that it promises (and often—but *not* always—produces). We examine where we are today and the challenge of tomorrow as the program moves into its next stages.

The alert reader will no doubt notice a modest amount of material repeated in several papers. To keep each paper complete in and of itself, I elected not to remove these duplications or to refer the reader to a section of another paper, feeling that would interrupt the flow. I hope the reader will agree with this editorial judgment.

ITS is part of the transportation system, which is a Complex, Large-Scale, Integrated, Open System (CLIOS), as described in my text *Introduction to Transportation Systems* (Artech House, Boston and London, 2000), and as discussed in "Transitions in the World of Transportation" (*Transportation Quarterly*, Vol. 56, No. 1, Winter 2002, Eno Transportation Foundation, Washington, DC, 2002) included in this volume. Transportation, as a CLIOS with a critical technological component, qualifies as an "engineering system", which implies that for its effective use, we need to go beyond the technology, and understand transportation and ITS on political, social, institutional, organizational and economic dimensions as well. That is not to downplay the technology—it is fundamental to transportation and ITS; technology enables both. But we do need to understand ITS in a more comprehensive way—on the dimensions of technology, systems and institutions—if we are to accomplish effective deployment and continued advances. I hope that this book is a step toward this broader understanding.

Acknowledgments

Over the past 16 years, I have had a number of rich experiences in the ITS world, all of which have contributed to this book. My seminal experience was participating in the development of the ITS (then IVHS) Strategic Plan while on sabbatical from MIT in 1991–92. It was at the invitation of Dr. James Costantino, then (founding) President of ITS America, that I had that opportunity. I will always be grateful to Jim for the career-altering opportunity he gave me. Jim, Thomas Deen, then Executive Director of the Transportation Research Board (TRB), Lyle Saxton, then of FHWA, and William Spreitzer of General Motors, served as the review panel for that strategic plan. Tom Deen also chaired the Strategic Planning Committee in those early days. All have been major figures in the development of ITS, and in my development as well.

My partners on the core writing team for the strategic plan, with their respective affiliations at the time, were Jonathan Arlook, Navigation Technologies Corporation; Edward Greene, Electronics Division of the Ford Motor Company; Craig Roberts, IVHS America; and Michael Sheldrick, Etak, Inc. They contributed to a great professional experience for me.

I would like to acknowledge the Transportation Research Board, ITS America and the Eno Foundation for permitting the reprinting of articles in this book. Their positive attitude toward the compilation of this volume was very gratifying to me.

Several public officials deserve special acknowledgment. Dr. Christine Johnson, long-time head of the Joint Program Office at FHWA (1994–2002), has been continuously supportive of my work in ITS.

Matthew Edelman, whom I have known since his student days at MIT in the early 1970s, and now Executive Director of TRANSCOM, has been an astute critic of much of my ITS work and a major contributor to the field in his own right.

Since 1993 I have taught a class called Introduction to Intelligent Transportation Systems (ITS) here at MIT. Let me acknowledge the students in those classes who, through their incisive questioning, helped shape my views on ITS; students made suggestions for some of the columns contained in this book.

My professional career has been enriched by people at MIT and elsewhere too numerous to mention. I appreciate all the help I had along the way.

As with my earlier book, I owe a huge debt of gratitude to Jan Austin Scott. She worked with me over the years in first developing the papers on which this book is based, and then in compiling them into a coherent whole. Her language sense and her good taste greatly improved the text. Her efforts were vital in this enterprise.

My graduate students, Sgouris Sgouridis and Travis Dunn, were very helpful in the final preparation and checking of this manuscript.

While this book has been improved by the comments of many of my colleagues, any errors or lack of clarity is the fault and responsibility of the author.

Contents

SECTION I. INTRODUCTION TO INTELLIGENT TRANSPORTATION SYSTEMS (ITS)

"The old order changeth, yielding place to new."
-- Alfred, Lord Tennyson
in *The Death of Arthur*

Section I is introductory and contains only my article "ITS: A Short History and a Perspective on the Future". This article provides an introduction to the fundamental concepts of ITS, its implications for surface transportation and how it contrasts with other transportation innovations and investment programs such as the Interstate system. It lays out the institutional issues central to ITS deployment and an initial strategic vision for ITS.

The strategic vision for ITS, then, is as the integrator of transportation, communications and intermodalism on a regional scale.

For additional articles relating to this theme, also see Section V-3.

I. 1. ITS: A SHORT HISTORY AND A PERSPECTIVE ON THE FUTURE[1]

1. INTRODUCTION

An article[2] on the development of Intelligent Transportation Systems (ITS) in this 50-year anniversary of the *Transportation Quarterly* and the 75th anniversary of the Eno Foundation is particularly appropriate. William Phelps Eno can in many ways be viewed as the great-grandfather of ITS. Eno's work in traffic control in the early days of highway transportation set the stage for the use of today's modern technologies in addressing the same issues with which he was concerned: congestion reduction and safety. I am honored to present this paper in this noteworthy issue of *Transportation Quarterly*.

This article describes the dynamic but short history of Intelligent Transportation Systems (ITS). ITS combines high technology and improvements in information systems, communication, sensors, and advanced mathematical methods with the conventional world of surface transportation infrastructure. In addition to technological and systems issues, there are a variety of institutional issues that must be carefully addressed. Substantial leadership will be required to implement ITS as an integrator of transportation, communications and intermodalism on a regional scale.

---

[1] Reprinted with permission of the Eno Foundation. Sussman, Joseph M., "ITS: A Short History and a Perspective on the Future", *Transportation Quarterly*, (Special Issue on the occasion of the 75th Anniversary of the Eno Foundation), Eno Transportation Foundation, Inc., Lansdowne, VA, p. 115-125, December 1996.

[2] Parts of this article are drawn from, "Strategic Plan for IVHS in the United States", Report No.: IVHS-AMER-92-3, Prepared by IVHS AMERICA, May 20, 1992. The author was a member of the core writing team that produced that document.

2.   HISTORY AND BACKGROUND

In 1986, an informal group of academics, federal and state transportation officials, and representatives of the private sector began to meet to discuss the future of the surface transportation system in the United States. These meetings were motivated by several key factors.

First, the group was looking ahead to 1991 when a new federal transportation bill was scheduled to be enacted. It was envisioned that this 1991 transportation bill would be the first one in the post-Interstate era. The Interstate System, a $130 billion program, had been the centerpiece of the highway program in the United States since the mid-1950s. By 1991 this project would be largely complete. A new vision for the transportation system in the United States needed to be developed.

While the Interstate had had a major and largely positive impact in providing unprecedented mobility at a national level, transportation problems remained. From the perspective of 1986, highway traffic delays were substantial and growing. Rush-hour conditions in many metropolitan areas often extended throughout the day. Further, safety problems abounded, particularly highway safety.

Also, the United States was concerned with the environmental impacts of transportation and the energy implications of various transportation policies. Any new initiatives in the surface transportation world had to explicitly consider environmental and energy issues.

Two more major motivations for considering the future of surface transportation were national productivity and international competitiveness, both closely linked to the efficiency of our transportation system. In 1986, our major economic rivals in western Europe (Project Prometheus) and Japan (Project AMTICS and RACS) were advancing very quickly in developing new technologies for use in advanced surface transportation systems. Their use of high technology concepts in the information systems and communications areas were seen as the opportunities to revolutionize the world of surface transportation. This would improve the competitiveness of these nations and provide them with an important new set of industries and markets.

Further, it was recognized that these congestion, safety, environmental and productivity issues would have to be addressed largely by means other than simply constructing additional conventional highways. Particularly in urban areas, the economic, social and political costs of doing so were becoming too high.

Thus, in 1986, this small, informal group saw before it an opportunity and a challenge based on:

• new transportation legislation (at that time five years in the future),

- concern for continuing transportation problems in the United States despite major investment in the transportation system,
- the development by our economic competitors in western Europe and Japan of various technologies that could enhance their industry posture and their productivity, and
- future limits on conventional highway construction, particularly in urban areas.

The essential concept was a simple one: marry the world of high technology and dramatic improvements with the world of conventional surface transportation infrastructure. The technological portion would include areas such as information systems, communications, sensors and advanced mathematical methods. This marriage could provide capacity with technological advances that could no longer be provided with concrete and steel. It could improve safety through technology enhancements and better understanding of human factors. Additionally, it would be able to provide transportation choices and control transportation system operations through advanced operations research and systems analysis methods.

What was envisioned and what came to be called Intelligent Vehicle Highway Systems (IVHS) and eventually Intelligent Transportation Systems (ITS), is but another example of the marriage of transportation and technology as a phenomenon that has existed throughout human history. In the early part of this century, innovation in construction and manufacturing technologies made the current transportation system possible.

We now have the need for a new round of technological innovation, appropriate to the transportation issues of today. For example, there are the "ITS-4" technologies. These technologies deal with:

1. the ability to *sense* the presence and identity of vehicles or shipments in real-time on the infrastructure through roadside devices or Global Positioning Systems (GPS);
2. the ability to *communicate* (i.e., transmit) large amounts of information more cheaply and reliably;
3. the ability to *process* large amounts of information through advanced information technology; and
4. the ability to use this information properly and in real-time in order to achieve better transportation network operations.

We use *algorithms* and *mathematical methods* to develop strategies for network control. These technologies allow us to think about an infrastructure/vehicle *system*, rather than independent components.

The small, informal group described above became "Mobility 2000", which produced a landmark document in 1990[3], laying out a vision for ITS.

---

[3] *Proceedings of a National Workshop on Intelligent Vehicle/Highway Systems Sponsored by Mobility 2000*, Dallas, TX, 1990.

In 1990, the need for a permanent organization became clear and IVHS America (the Intelligent Vehicle Highway Society of America) was formed as a federal advisory committee for the U.S. Department of Transportation.

In December 1991, the Intermodal Surface Transportation Efficiency Act (ISTEA) became law. Its purpose was "...to develop a National Intermodal Transportation System that is economically sound, provides the foundation for the nation to compete in the global economy, and will move people and goods in an energy-efficient manner."

As was envisioned in 1986, ITS was an integral part of ISTEA, with $660 million allocated for research, development and operational tests. Additional federal, state, local and private-sector funds were added to this initial allocation, leading to a substantial program.

In June 1992, IVHS America produced "A Strategic Plan for Intelligent Vehicle Highway Systems in the United States" and delivered it to the U.S. DOT as a 20-year blueprint for ITS research, development, operational testing and deployment.

The vision for ITS was articulated as:

- A national system that operates consistently and efficiently across the United States to promote the safe, orderly and expeditious movement of people and freight. Here, recognition of the need to think intermodally and about the needs for *both* personal and freight mobility was explicit.
- An efficient public transportation system that interacts smoothly with improved highway operations. The concept that ITS had to do more than simply improve single-occupancy-vehicle level-of-service on highways is captured here.
- A vigorous U.S. ITS industry supplying both domestic and international needs. The plan noted that the U.S. market for ITS hardware and software services would be on the order of $230 billion over the next 20 years. Extrapolating this internationally, it is not unreasonable to think about a $1 trillion international market in ITS over that time period, well worth the effort for the private sector to pursue.

The strategic plan talked about a vision for ITS, the mission for the ITS community, and the setting in which ITS would be deployed. It addressed strengths, weaknesses, opportunities and threats to developing and deploying ITS in the United States. Specific goals and objectives, as well as a formal plan for research, development, testing and deployment were outlined in the plan. This detailed plan included a discussion of the fundamental approach to ITS, the research and development necessary to reach the deployment stage, systems integration, and estimates of funding needed for the program.

The plan focused on three facets: First, *technology*, the development and integration of technologies that would allow ITS to proceed. Second, *systems*, the integration of technologies into systems for operating ITS. Third, *institutions*, the challenges that face the ITS community in developing the public-private partnership and the government interactions that would have to be developed at various levels. This third facet also addressed the educational challenge that the community faces and the organizational changes that would be necessary to have success in the ITS theater. It was recognized that while technology and systems were important issues in the development of ITS, the many institutional and organizational issues would be as complex and as difficult.

In the Strategic Plan, the recognition of *the transportation/ information infrastructure* was an important conceptual breakthrough. In other words, it accepted an intermingling of the new technologies in computers, communications and sensors, with conventional infrastructure to create something wholly new in the world of transportation. In thinking about the development of this concept, it is helpful to use a construct developed by one of the seminal thinkers in management, Peter Drucker. He talks about three revolutions that have occurred in the last 200 years: the industrial revolution, the productivity revolution and the management revolution. In the industrial revolution, roughly 200 years ago, new technology replaced manpower with machine power, with extraordinary improvements in capacity.

About 100 years ago, we experienced the productivity revolution. This involved applying *knowledge to work*. It included such concepts as the assembly line and ideas in scientific management developed by Frederick Taylor. This productivity revolution has given rise to extraordinary improvements in productivity (about 4.5% annually over a long time period) and great improvements in quality of life. However, Drucker argues that this revolution is largely over. It applies mainly to improving the productivity of manual workers, who, in the developed world of 1950, represented about half the workers. In 1990, only about 20% of the workers, and in 2010, perhaps 10% of the workers are, or will be, manual laborers. There is only modest leverage left here.

Drucker argues that what is needed is a new revolution he calls the management revolution which is the idea of applying *knowledge* not to work but *to knowledge*. This will allow us to greatly enhance the productivity of non-manual workers. It can also be a mechanism for creating wealth and quality of life at a level we have been unable to achieve through the industrial and productivity revolutions over the last 200 years.

ITS is part of this management revolution. A systemic approach linking vehicles and infrastructure is something fundamentally new in transportation.

It is convenient to think of ITS in terms of six areas: Advanced Traffic Management Systems (ATMS), Advanced Traveler Information Systems (ATIS), Advanced Vehicle Control Systems (AVCS), Commercial Vehicle Operations (CVO), Advanced Public Transportation Systems (APTS), and Advanced Rural Transportation Systems (ARTS).

3.   FUNCTIONAL AREAS IN ITS

3.1   Advanced Traffic Management Systems (ATMS)

ATMS will integrate the management of various roadway functions. It will predict traffic congestion and provide alternative routing instructions to vehicles over regional areas to improve the efficiency of the highway network and maintain priorities for high-occupancy vehicles. Real-time data will be collected, utilized and disseminated by ATMS and will further alert transit operators of alternative routes to improve transit operations. Dynamic traffic control systems will respond in real-time to changing conditions across different jurisdictions (i.e., by routing drivers around accidents).   Incident detection will be a critical function in reducing congestion on the nation's highways.

3.2   Advanced Traveler Information Systems (ATIS)

ATIS will provide data to travelers in their vehicles, in their homes or at their places of work. Information will include: location of incidents, weather problems, road conditions, optimal routings, lane restrictions, and in-vehicle signing. Information can be provided both to drivers and to transit users and even to people before a trip to help them decide what mode they should use.

3.3   Advanced Vehicle Control Systems (AVCS)

AVCS is viewed as an enhancement of the driver's control of the vehicle to make travel both safer and more efficient. AVCS includes a broad range of concepts that will become operational on different time scales.

In the near term, collision warning systems would alert the driver to a possible imminent collision. In more advanced systems, the vehicle would automatically brake or steer away from a collision.   Both systems are autonomous to the vehicle and can provide substantial benefits by improving safety and reducing accident-induced congestion.

In the longer term, AVCS concepts would rely more heavily on infrastructure information and control that could produce improvements in roadway throughput of five to ten times.   This concept is called the Automated Highway System (AHS). Movements of all vehicles in special lanes would be automatically controlled. One could envision cars running in closely-spaced (headways of less than one foot) platoons of ten or more, at normal highway speed, under automatic control.

ATMS and ATIS will have early applications in urban and suburban areas. AVCS, particularly the Automated Highway System (AHS), is envisioned as a longer-term program. In addition, CVO, APTS and ARTS are three major applications areas that are already beginning to draw on ITS technologies.

3.4   Commercial Vehicle Operations (CVO)

In CVO, the private operators of trucks, vans and taxis have already begun to adopt ITS technologies to improve the productivity of their fleets and the efficiency of their operations. This is proving to be a leading-edge application because of direct, bottom-line advantages.

3.5   Advanced Public Transportation Systems (APTS)

APTS can use the above technologies to greatly enhance the accessibility of information to users of public transportation as well as to improve scheduling of public transportation vehicles and the utilization of bus fleets.

3.6   Advanced Rural Transportation Systems (ARTS)

The special economic constraints of relatively low-density roads and the question of how ITS technologies can be applied in this environment are challenges that are being undertaken by many rural states.

4.   A BROAD APPROACH

ITS represents a broad systemic approach to transportation. ATMS represents overall network management. ATIS is the provision of information to travelers. AVCS is a new level of control technology applied to vehicles and infrastructure. Applications in urban and rural areas, involving public transportation, commercial vehicles and personal highway vehicles, are encompassed by ITS.

There are important technological issues to be considered, many involving the *integration* of various hardware and software concepts on a "real-world" transportation network. Few technological "breakthroughs" (with the likely exception of AHS) will be needed.

5.   INSTITUTIONAL ISSUES

Of equal importance to technological and systems issues are various *institutional* issues that must be addressed if ITS is to be successfully deployed. Several are discussed below.

5.1   Public-Private Partnerships

A primary issue is the need for public-private partnerships for ITS deployment. One can contrast ITS with the Interstate System, the major transportation program in this nation in the 20th century. The Interstate System could be characterized as a public works system. The funding was provided exclusively by the public sector and the fundamental decisions about the deployment of the Interstate System were made by the public sector.

ITS, on the other hand, will require deployment of infrastructure, largely by the public sector, and in-vehicle equipment by the private sector. Therefore, ITS can be characterized as both a *public works and* a *consumer product* system. This will require unprecedented levels of cooperation between the public and private sectors if ITS is to work effectively as a national "seamless" system. The hardware and software in the infrastructure must be compatible with the hardware and software in the privately-acquired in-vehicle equipment.

While stand-alone ATMS (i.e., infrastructure) and ATIS (i.e., in-vehicle equipment) could work well, researchers are convinced that coordinated use of ATMS and ATIS will be much more effective than stand-alone systems of either type. Therefore, for optimal system operations, coordination and compatibility between ATMS and ATIS is essential. This requires close cooperation between the public and private sectors. In the United States, this cooperation has often not been strong. So ITS presents an important set of institutional challenges in developing an effective public/private partnership for ITS research and development, testing and deployment.

5.2   Organizational Change

A second institutional question is the need for organizational change brought about by ITS. For example, our state Departments of Transportation have been based, for many decades, upon the technology of traditional civil engineering. Highway construction and maintenance have been the charter of state DOTs and, in fact, they have built a highway system that is unrivaled in the world.

However, that world is changing with socio/political/economic constraints and with ITS coming on the scene. Now, rather than dealing with the conventional civil engineering technologies of structures, materials, geotechnical engineering and project management, state DOTs need to be concerned with electronics, information systems, communications and sensors. DOTs will need to emphasize the operation of the transportation system as well as construction and maintenance.

This is a fundamental shift for these public organizations. They will have to make a difficult transition over the next several decades for ITS to be successfully deployed around this nation, as will private-sector organizations that have supported the historical mission. A whole new set of professionals will need to be attracted to these public-sector organizations and related private-sector organizations. In addition, fundamental changes in the mission of these organizations must come about.

It is interesting to observe that on the Central Artery/Tunnel program in Boston, Massachusetts, one of the last major projects of the Interstate System, ITS is playing a major role. The contractor on the project is putting considerable resources into understanding and developing ITS systems that can be used in conjunction with the development of conventional infrastructure to make sure this mammoth ($10 billion) megaproject will, in fact, work. Together with the Commonwealth of Massachusetts, the Massachusetts Institute of Technology and MIT's Lincoln Lab, the contractor is working on traffic control centers, algorithms for effective routing of traffic, and roadside infrastructure that will permit efficient monitoring of traffic and incident detection.

The symbolism is strong. One of the great international construction consortiums working on the last of the great Interstate projects in this country is focusing on ITS technology to enable the finished project to operate effectively.

5.3   Academia in Transportation

The development and deployment of ITS imply important issues for academia, both in research and in the education of new transportation practitioners. The academic community has a major role to play and has already seen an opportunity in ITS as a number of active programs have already been initiated. For example, the University of California, University of Michigan, University of Minnesota, University of Texas, Texas A&M, MIT and others have active ITS research and education programs.

The most important function of academia is the development of educational programs and the education of transportation professionals. The deployment of ITS implies change. for transportation organizations. Consequently, a broader education of the transportation professional, including areas such as software systems, communications, a variety of systems analysis and operations research methodologies, information systems, and institutional studies will be required. What is needed is a "new synthesis" as an educational model for the "new transportation professional".[4] The development of that synthesis and the education of the new transportation professional will be a critical contribution by the academic community.

Academia also has an important role to play in research activities in the ITS arena. Academia will be a major participant, both in assessing the current state of likely technological improvements and in providing basic and applied research and development.

Indeed, there is a close tie between the research programs and the new synthesis noted above. ITS research will require the talents of faculty in areas that have not traditionally been involved in transportation. The access to interesting research problems, as well as to funding, provides the pathway and motivation for new faculty to participate in transportation research in the university. It will be essential to engage those faculty members in transportation education and developing the new synthesis. That approach has worked effectively in fields such as manufacturing and biomedical engineering, and ITS is an opportunity to make it happen in transportation as well.

Success in ITS will require progress in three areas: the "triad" of technology, systems and institutions and management, mentioned earlier in the context of the ITS strategic plan. The development and integration of advanced technology into the transportation infrastructure is central to ITS. Systems level activities, including network operation, economic analysis, optimization and simulation are likewise fundamental. Finally, institutional and management issues such as public/private partnerships, intergovernmental relations and legal questions are also of prime importance.

These three areas require a breadth of capabilities not captured by many organizations. The modern research university is best suited for such broad activities. These universities, with their dual roles in education and research, have built broad faculties in engineering, management, political science and technology policy. They often undertake mission-oriented work that requires the broad vision and expertise described above. By addressing the

---

[4]   Sussman, Joseph M., "Educating the 'New Transportation Professional,'" *ITS Quarterly*, ITS America, Washington, DC, Summer 1995. N.B. This article appears in Section III of this volume.

triad in an effective way, research universities have a unique role to play in the ITS arena.

6.　　TRANSPORTATION AND CHANGE

The linking of conventional infrastructure with the technologies of information systems, communications, sensors and advanced mathematical methods for the movement of both people and freight is an extraordinary development. We cannot begin to foresee the changes (possibly both positive and negative) that will result from the development of this transportation/information infrastructure.

Think, for example, about the changes that came about as a result of the Interstate System, a $130 billion program, starting in 1956. The Interstate program can be thought of as an expansion, *in-kind*, of a conventional highway system. Granted, the Interstate was a substantial expansion in capacity and network size, but it was an in-kind improvement nonetheless. Yet, we had a hard time predicting what would happen as a result of this implementation. For example:

- The inter-city trucking industry was formed, and a financial blow was dealt to the railroad industry, as it lost substantial market in high-value freight. This led, in turn, to a fundamental redefinition of the relationship between the public and private sectors in the freight industry in 1980, through substantial deregulation.
- The Interstate led to an unprecedented and unequaled mobility between and into U.S. cities and gave rise to the regional transportation concept, with wholly new methods of planning being required for region-wide analysis and design.
- The Interstate System included the development of circumferential belts around major cities, leading to development patterns quite at variance with the ability of public transportation to service it and, as described by authors such as Joel Garreau, the development of "edge cities", a fundamentally new kind of urban structure.
- The Interstate led to a fueling of the post-war economic expansion and a period of unprecedented prosperity in the United States.
- A "stop the highway" backlash in urban areas resulted from the Interstate, and a political polarization between the build vs. no-build factions became a fact of political life in U.S. transportation.

All of this results from an expansion, *in-kind*, of the highway system.

Regarding ITS, we have already seen:

- the reinvention of logistics through supply chain management, linking inventory management and transportation in wholly new ways;

- dramatic moves into surface transportation by organizations not traditionally involved, such as the national labs and aerospace companies in the United States;
- changes to academia, with new alliances and new academic programs beginning to form and faculty participating in transportation education and research who have never been part of that process before; and
- the building of new relationships among public-sector agencies to enable regional and corridor-level system deployment.

These have already happened, and it is just the beginning. We cannot begin to foresee all that will occur. The enabling technology of ITS, the transportation/information infrastructure, can and will have profound effects. We hope they will be positive -- accessibility, economic growth, improved quality of life, improved information for planning and intermodal transport. However, unforeseen outcomes, both positive and negative, are certain with this new transportation enterprise.

We need to think broadly about ITS and the transportation/information infrastructure. Our job is the mobility of people and goods and, in fact, knowledge, as Peter Drucker would put it. ITS is an enabling technology to permit us to do great things in these areas. We cannot foresee all the possibilities of ITS -- all its potentials and all its pitfalls. Still, it is certainly the most exciting development in transportation in many years. We, as transportation professionals, need to make it work.

7.   THE POST-STRATEGIC PLAN PERIOD

The years since the Strategic Plan have been busy ones in the ITS community. Program plans which translate the Strategic Plan into specific shorter-term actions have been developed. A national ITS system architecture has been developed. The U.S. DOT has established the Joint Program Office as a group that cuts across the modal administrations of DOT to address ITS research, development, testing and deployment. ITS America continues to grow, with 1,000 members in 1996, and more than 3,000 people at the most recent ITS annual meeting in Houston, Texas. The international community in ITS is cooperating at a professional level with the ITS World Congress, initiated in Paris in 1994, continued in Yokohama in 1995, and being hosted in the United States in Orlando, Florida, in 1996.

Space does not permit documenting all the ITS successes to date. To name but several: One can look at TRANSCOM in the congested New York/New Jersey/Connecticut region for an example of an ITS deployment providing ATMS, ATIS and electronic toll collection in the tri-state area. The SmarTraveler program in Boston is an example of an advanced traveler information system with a strong initial track record. The Houston public transportation system is yet another example of an ITS deployment which is quantitatively and qualitatively changing the supply of transportation service in the Houston metropolitan area. Various deployments in Western Europe and Japan are advancing as well.

In 1996, at the Transportation Research Board Annual Meeting in Washington, Secretary of Transportation Federico Peña announced the Intelligent Transportation Infrastructure (ITI) and Operation Timesaver, with ambitious goals for deploying ITS technologies throughout the country by early in the 21st century.

Private-sector organizations have continued active programs in ITS. The automobile manufacturers in the United States and abroad are marketing advanced traveler information systems, supported by major initiatives in communications and in computerized mapping.

Cooperation between various public-sector agencies also characterizes the ITS movement. Commercial vehicle operation initiatives, such as HELP and Advantage I-75, involve many public jurisdictions as well as private-sector truckers. The I-95 coalition cooperates on ITS technologies stretching from New England to Virginia.

Efforts to make ITS truly intermodal, both for travelers and freight, are underway. For example, exploiting ITS to enhance truck-rail-ocean freight intermodalism is high on the agenda.

ITS technologies and concepts have begun to be embedded within the transportation system. The benefits of such technologies are being established through the collection of field data that supports the claims of congestion reduction and safety enhancement. A new round of advances and deployments is imminent.

The reauthorization of ISTEA, the 1991 federal legislation that first enabled the ITS program, scheduled for 1997, will be a major milestone. This should happen as the ITS community anticipates its major role within this legislation to move to the next level of system deployment.

8.   CONCLUSION

The focus for ITS in the future is clearly on deployment. Taking research and operational test results and putting them into routine practice is the emphasis in the ITS world today.

How to best advance the deployment agenda is currently a matter of intense discussion in the ITS community.

The best approach is for ITS to focus on regions as critical units of economic competition.[5] Often, we speak of the "competitive region". The work of Professors Michael Porter and Rosabeth Moss Kanter at the Harvard Business School emphasizes the idea that subnational units will compete economically on the basis of productivity and quality of life provided for its citizens.

This concept can be combined with two others. First, the natural partnership between ITS and the nascent National Information Infrastructure (NII), a communications network of unprecedented scale, scope and functionality, can provide substantial deployment benefits to both.

Second, the strong trend toward freight and traveler intermodalism provides a critical boost to ITS technologies. This is where ITS can help overcome intermodalism's weak point -- the transfer process -- through information and communication technology.

Pulling these ideas together:

> The strategic vision for ITS, then, is as the integrator
> of transportation, communications and intermodalism on
> a regional scale.

This is an ambitious vision and one that will require substantial leadership to achieve the technology deployment and the institutional change that will be needed to achieve such an outcome.

ITS has had a dynamic but short history. Challenges have been overcome in these early years. Many remain for the future. When the *Transportation Quarterly* celebrates its 100th anniversary, I believe we will look back on this time as a seminal period in the history of transportation and one in which a truly intermodal transportation/information infrastructure was deployed, advanced by our ITS program.

---

[5]   Sussman, Joseph M., "ITS Deployment and the 'Competitive Region,'" *ITS Quarterly*, ITS America, Washington, DC, Spring 1996. N.B. This article appears in Section II of this volume.

Bibliography

1. Drucker, Peter, <u>The Age of Discontinuity: Guidelines to Our Changing Society</u>, Harper & Row, New York, 1969.
2. GAO Report, *Smart Highways — An Assessment of Their Potential to Improve Travel,* May 1991.
3. *A Strategic Plan for IVHS in the United States*, ITS America Report No. IVHS-AMER-92-3, May 1992.
4. Sussman, Joseph M., "Educating the 'New Transportation Professional,'" *ITS Quarterly*, ITS America, Washington, DC, Summer 1995.
5. Sussman, Joseph M., "Intelligent Vehicle Highway Systems and the Construction Industry," *Construction Business Review*, May/June 1993.
6. Sussman, Joseph M., "ITS Deployment and the 'Competitive Region,'" *ITS Quarterly*, ITS America, Washington, DC, Spring 1996.
7. Transportation Research Board, *Advanced Vehicle and Highway Technologies, Special Report 232,* December 1991.
8. U.S. DOT, *Moving America -- New Direction, New Opportunities: A Statement of National Transportation Policy,* February 1990.

SECTION II. ITS ORGANIZATIONAL ISSUES, REGIONALISM AND THE TRANSPORTATION/INFORMATION INFRASTRUCTURE

"Systems thinking produces radical and potentially revolutionary visions of public institutions. Nothing short of such visions can transform the state of world affairs."

-- Russell Ackoff

Three central ITS ideas -- all of which are new to the world of transportation and of fundamental strategic importance -- are considered and integrated in this section.

1. ITS requires important organizational and institutional change in the transportation world. How ITS relates to existing organizations in the transportation field and how new organizations must evolve to meet the challenge of ITS deployments is an issue well worth discussing.
2. ITS technologies enable the operation and management of transportation systems at a *regional scale*, usually for the metropolitan-based region. This new scale is of vital importance in the world of surface transportation. This creates opportunities for more effective transportation operations, but to achieve this improved effectiveness we need to overcome some formidable institutional barriers to regionalization.
3. ITS defines a new kind of infrastructure -- the *transportation/ information infrastructure*. This is the merging of the physical movement of the people and goods around a transportation network -- that is, transport -- and communication of information regarding and germane to those movements.

So, institutional change to deal with the regional opportunities created by ITS through the advent of the transportation/information infrastructure is the integrating theme of this section.

For additional articles relating to these themes, also see Section IV-1, 2, 10.

II. 1. TRANSPORTATION OPERATIONS: AN ORGANIZATIONAL AND INSTITUTIONAL PERSPECTIVE[1]

1. INTRODUCTION

The essence of this article is simple. Because of the need for customer and market focus in providing surface transportation, and constraints on building conventional infrastructure, the emphasis in modern surface transportation systems must be on *operations*, enabled by new advanced technologies. This operations focus, together with the new technologies, in turn, requires change to transportation organizations dealing with what to many is a new mission. We argue that operations are most appropriately and effectively carried out at the regional scale, with information-sharing and responsibility-sharing among these changed organizations. This, together with changes in funding patterns reflecting shifts from capital to operations expenditures, requires institutional change in the relationships among these organizations. Institutional changes at all levels of government -- federal, state, regional and local -- and the private sector are a *required precondition* for an operations, customer-oriented focus. These relationships are captured in the following diagram.

---
[1] This article originally appeared as Sussman, J. M. *Transportation Operations: An Organizational and Institutional Perspective*. Report for National Special Steering Committee for Transportation Operations and Federal Highway Administration/U.S. Department of Transportation, http://www.ite.org/NationalSummit/index.htm., Washington, DC, December 2001.

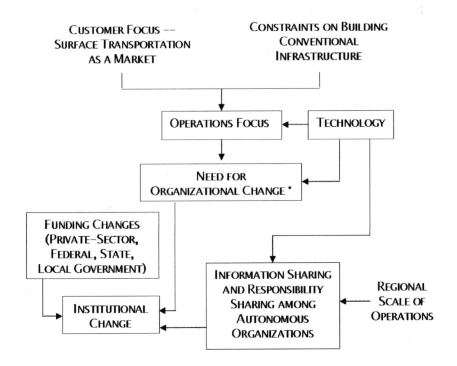

* E.G.,   HUMAN RESOURCE DEVELOPMENT, REWARD AND INCENTIVE STRUCTURE
          FOR OPERATIONS-ORIENTED PROFESSIONALS

This article focuses on the specifics of the organizational and institutional change required by the emerging focus on transportation operations, and some ideas about how that can be achieved in practice.

The overarching context for this article is the re-authorization of TEA-21, currently several years in the future. This omnibus transportation bill has historically set the tone for transportation investment, priorities and institutional change, as witness ISTEA in 1991 and TEA-21 in 1997. The intent of this article is to identify and discuss important ideas, relating to transportation operations and the associated institutional and funding changes, so as to contribute to the informed debate leading up to TEA-21 re-authorization. Certainly a key aspect of this debate is the notion of congestion relief as a *federal* responsibility. Orski notes that, while

> "There is much to be said against "federalizing" every new problem that confronts the nation. Recent experience, however, suggests that when travel delays reach an

unacceptable level, Congress will not hesitate to intervene."[2]

Operations as a mission relates to other important potential policy initiatives within TEA-21 re-authorization.   One is the concept of *sustainable transportation* -- developing a transportation system which provides for the mobility needs of our people, while at the same time, avoiding critical negative environmental impacts. Sustainable transportation is defined as follows:

A sustainable transportation system is one that:
- *allows the basic access needs of individuals and societies to be met safely and in a manner consistent with human and ecosystem health, and with equity within and between generations.*
- *is affordable, operates efficiently, offers choice of transport mode, and supports a vibrant economy.*
- *limits emissions and waste within the planet's ability to absorb them, minimizes consumption of non-renewable resources, reuses and recycles its components, and minimizes the use of land and the production of noise.*[3]

We suggest that the operations focus will contribute in substantial ways to a sustainable transportation system.

This refocus of the transportation enterprise first toward emphasizing operations and not simply a sole emphasis on conventional infrastructure, and second toward a dual focus on mobility and sustainable transportation and away from a sole focus on mobility, will require substantial organizational and institutional changes. We recognize at the outset that organizational and institutional change is inherently difficult.   Yet it is essential if the operations mission before us is to be effectively addressed. The purpose of this article is to identify and describe important issues vis-à-vis this organizational and institutional change and suggest ways to create it.   We deal with the practicing professional and the organizations within which s/he works, and the relationships among those organizations. These organizational and institutional changes are strategic in nature. To accomplish them we need to understand the barriers to such change. This article, building on the work of many authors in the field, discusses what those barriers are and what can be done to overcome them.

---

[2]   Orski, Kenneth, "Congestion Relief Should Become an Explicit Objective of the Federal Surface Transportation Program", *Innovation Briefs*, Vol. 12, No. 5, Sep/Oct 2001.

[3]   The Centre for Sustainable Transportation. *A Definition and Vision of Sustainable Transportation*, July 2001. http://www.cstctd.org/CSTmissionstatement.htm.

We begin by explaining why emphasis on transportation operations is important and what drives that emphasis. The reader is invited to read Appendix C for detailed definitions of various terms.

2. DRIVING FORCES TOWARD AN OPERATIONS FOCUS[4]

The emerging emphasis on transportation operations (as opposed to capital investment in transportation facilities) is driven by several important factors.

The transportation world is increasingly customer-driven. Following the lead of the Internet society, our transportation systems must increasingly take a customer perspective. The days of "one size fits all" in provision of transportation service is fading. Surface transportation must be viewed as a market with a heterogeneous customer mix. An early example is HOT lanes, where customers willing to pay a premium price for the use of highway infrastructure, do so and receive a premium service. The customer orientation *requires* an operations perspective.

What Our Customers Need

The needs of traveler and freight customers is our overarching concern. From a strategic point of view, our customers are concerned with quality of life from an economic and environmental perspective, and with sustainable economic development. Concern with safety and security is also a primary customer need. Safety *is* an operating question, as identified by Olmstead[5] when he relates the operation of variable message signs and traveler information systems as statistically linked to positive changes in safety performance on highways. Security, highlighted by September 11, 2001's tragic events, will doubtless be of increasing concern.

From a tactical point of view, our customers are concerned with mobility and accessibility. Our customers want transportation choices and real-time information about

---

[4] This section draws upon Sussman, Joseph M., David Hensing and Douglas Wiersig, "Transportation Operations and the Imperative for Institutional Change", Working Paper for ITE Annual Meeting, Irvine, CA, April 2000.

[5] Olmstead, Todd, "The Effects of Freeway Management Systems and Motorist Assistance Patrols on the Frequency of Reported Motor Vehicle Crashes", Doctoral Thesis, Harvard University, Cambridge, MA, May 2000.

those choices. Improved travel time and congestion relief have been identified by our customer base. More subtly, customers desire a minimization of *unpredictable* delays, emphasizing the "*reliability*" of the transportation enterprise. Customers may be willing to live with longer travel times; they find it more difficult to live with high variability of those travel times on a day-to-day basis. This concept has long been understood on the freight operations side. Research on service reliability in the rail industry goes back many decades.[6] It is relatively recently, though, that performance measures for highway systems have begun to incorporate this day-to-day variability or unreliability in traffic operations.[7]

The increased focus on operations also results from the limits of our ability to provide conventional infrastructure, particularly in urban areas. Here the social, political, economic and environmental forces mitigate against our traditional "build our way out of it" approach to providing transportation capacity. Fortunately, more effective operations provide an alternative path.

The focus on operations is enabled by a set of new technologies -- especially Intelligent Transportation Systems (ITS). These technologies, which permit an electronic linkage between vehicle and infrastructure, create an environment in which management of transportation operations can take a major leap forward. These technologies also drive the need for organizational and institutional change.[8]

ITS and Operations

While ITS provides important technologies to support transportation operations, ITS and operations are not identical concepts. There are operational issues which have little or no tie to the technologies of ITS. On the other hand, some components of ITS relate more to planning than to operations.

---

[6]   Sussman, Joseph M. and Joseph Folk, "Unreliability in Railroad Network Operations", a chapter contributed to *Analytic Foundations of Engineering Problems: Case Studies in Systems Analysis*, Richard de Neufville and David Marks, eds., January 1973.

[7]   Lomax, Tim, Shawn Turner, Mark Hallenbeck, Catherine Boon, Richard Margiotta, "Traffic Congestion and Travel Reliability: How Bad Is the Situation and What Is Being Done about It?", July 2001.

[8]   See, for example, NCHRP Synthesis 296, Transportation Research Board, National Research Council, National Academy Press, Washington, DC, 2001.

Nonetheless, many of the organizational and institutional barriers that ITS faces are close to or identical to issues in the broader operations theatre. For example, the funding issues associated with ITS deployment are closely related to the difficulties public agencies have in obtaining support for continuing operations, as opposed to large-scale capital projects. So, in this paper, while recognizing that ITS and operations are not identical, we are still able to utilize the ITS experiences in building our understanding of the organizational and institutional barriers to the broader operations questions.

Operations can now be deployed at the scale of the metropolitan-based region. There is an increasing consensus that the unit of economic competition in our global economy is the metropolitan-based region and not the nation. Strategists such as Michael Porter and Rosabeth Moss Kanter from the Harvard Business School emphasize this regional perspective in their writings. Also, to be effective, environmental issues -- e.g., clean air and water, land use -- need to be addressed at the regional, not simply the urban, scale. The ITS technologies noted above can allow the transportation system to be managed at that same regional scale[9]. So a further driving force is our ability to operate and manage transportation systems at the same regional geographic scale at which economic competition and environmental concerns take place.

This change to an operations mission at a regional scale requires new approaches in *technology, systems* and *institutions*. In the author's view, we have the technology in hand to create effective operations. Our understanding of broad-based systems and their behavior is fast approaching the level of knowledge that we require for complex system operations. However, on the *organizational and institutional* dimension, *major change is required* to work at a regional scale because of the information- and responsibility-sharing inherent in this scale. And these kinds of changes are difficult to achieve.

Further, *institutional change* will be required by the different funding forms required by this operations focus. Operations requires continuing, reliable year-to-year funding to be successful, unlike the more front-loaded, one-time funding for capital projects. This continuing funding has usually been more difficult to obtain for public agencies and is often the first victim of cuts in difficult economic times. To achieve operating success, an

---

[9] Sussman, Joseph M., "ITS Deployment and the 'Competitive Region'", "Thoughts on ITS" Column, *ITS Quarterly*, ITS America, Washington, DC, Spring 1996. N.B. This article appears in this section (Section II) of this volume.

institutional structure which assures continuing and reliable funding will be needed.

We will need change and leadership at various levels. We will need regionally-scaled organizations to deal with transportation and related issues at that geographic scale. We will need changes at the federal level to create funding mechanisms for operations to these regionally-scaled organizations that, in turn, need mechanisms whereby they can effectively accept and disburse these funds. What is required is federal leadership of the sort that existed in the early days of the Federal Aid Highway Program and the development of the Interstate program decades later.[10] Leadership comparable to that provided by the states throughout the 20th century will also be needed to put together the regional coalitions so critical to the operations focus.

We know that mission change and new technology require organizational and institutional change. Witness what the mission change for the U.S. military since the end of the "Cold War" has required organizationally and institutionally. Witness what the advent of containerization technology has meant for the structure of the global freight system. These changes occur at a much slower pace than we see in technology or systems, but to be successful in building an operations perspective, we need to provide a professional and political environment that can expedite the adoption of technology and systems needed for the operations mission.[11]

3. THE CASE FOR ORGANIZATIONAL AND INSTITUTIONAL CHANGE

If one accepts the need for an operations mission at a regional scale around the United States, the need for organizational and institutional change to deliver on that mission is compelling. As effective as our state and local transportation organizations have been in delivering the infrastructure-intensive surface transportation systems of the last century, they are not designed, for the most part, to deliver on the operations mission for the 21st century. This section lays out the arguments, first by explaining

---

[10] Weingroff, Richard F., "For the Common Good: The 85th Anniversary of a Historic Partnership", *Public Roads*, March/April 2001.

[11] The transportation field is subject to many changes here in the early years of the 21st century. The article "Transitions in the World of Transportation: A Systems View" (MIT Engineering Systems Division Working Paper, June 2001), is based on a talk given by the author at the April 2001 ITS Summit held in Washington, DC. Its theme of "transitions" provides a basis for thinking about the organizational and institutional changes that must occur in the transportation field. N.B. A later version of this article appears in Section V of this volume.

why institutional issues occur, and then by focusing more explicitly on various organizational and institutional considerations.

3.1   Why Institutional Issues Arise[12]

At this point, it is useful to do a "first-order" identification of some fundamental reasons institutional issues arise. Among those reasons are:

- *Concern with autonomy.* Creating linkages among organizations and potentially creating new organizations, be they virtual or real, can lead to a loss of autonomy for the participating organizations. Those organizations may feel they are unable to discharge the responsibilities they were chartered to do if that autonomy loss occurs.
- *Mission mismatch.* Different organizations such as state DOTs, MPOs, law enforcement agencies, and so forth, have different core missions. The missions may, in fact, be complementary, but the different mindsets these organizations bring to the table may cause institutional difficulties.
- *Differences in resources.* Budgets may be different in various jurisdictions, leading to difficulty in all organizations being able to perform as equal partners.
- *Funding sources.* Institutional issues will occur if funding sources are not consistent with the organization's mission. If traditional funding sources are directed to, say, capital spending and an additional mission focuses on operations, that disconnect generates an important institutional issue between funder and fundee.
- *Ideology.* Noted earlier in this article is the idea of considering surface transportation as a market with differentiated service and prices for customers with different needs and willingness to pay. This point has ideological content, particularly in an environment in which a traditionally public service -- highway and public transportation -- is being offered. Such a conceptual change to basic principles will certainly generate institutional concerns.
- *Technology.* Different organizations take different technological approaches to meet their missions. This may lead to difficulties in making technical systems interface properly. Further, these organizations may have different staff capabilities in technical areas, making sharing responsibilities equitably difficult.
- *Information.* The operations mission runs on information. There may be concern among various organizations about sharing that information, and in some cases there may be difficulties (or reluctance) for some

---

[12] Some of the extensive literature on institutions and operations in transportation is reviewed in Appendix A. The companion papers developed in parallel to this one are reviewed in Appendix B.

organizations in delivering the necessary information to their partners. Integrating information may present a difficult technical problem.

The City of Bangkok has long been plagued by major traffic problems, since the development of wealth in that country led to dramatic rates in growth of automobile ownership that far outstrip the ability of conventional infrastructure to serve it. In one attempt to deal with traffic issues, the Bangkok Traffic Department deployed SCAT, an Australian-developed traffic management system for its traffic signals in downtown Bangkok. SCAT attempts to globally optimize vehicular traffic by setting traffic signals, changing cycle times, red and green splits, and so forth.

For political reasons, and presumably in response to emergency-preparedness requirements, individual police officers on various street corners were permitted, in times of "emergency", to take the signals at their intersection off central control and manage their intersection manually. What actually happened was that police officers, with some pride in the operation of "their" street corner, would remove the signals not only at times of emergency, but when congestion appeared to them to be "excessive" (which could be rather often). And, indeed, these police officers would often be successful in clearing their intersection; however, they would wreak havoc on the traffic system at-large. Central management of traffic could not cope with individual police officers taking their intersection into their own hands, however well-meaning that tactic might have been.

The lessons:
1. This is an institutional issue -- in this case, between the Bangkok Police Department and the Bangkok Traffic Department -- as they both worked with good intentions toward curbing the congestion beast in one of the world's biggest and most congested megacities.
2. Suboptimization can be destructive of even the most sophisticated centrally-controlled systems. People can trump technology if not properly instructed.
3. Operations requires *discipline*. While there may be circumstances in which a true emergency warrants putting an intersection on manual control, this is a rare event. Mere congestion is not such an event.

Institutional issues arise, even in straightforward situations. In Appendix A, this author discusses a paper by James A. Powell[13]. Mr. Powell's paper is quite interesting, albeit discouraging. He describes the need to coordinate among three major metropolitan areas -- Gary, Indiana, Chicago, Illinois, and Milwaukee, Wisconsin -- and other transportation organizations on a corridor project. He identifies 41 coordination issues, most of which struck this author as quite straightforward; yet the institutional difficulties in getting them resolved was extraordinary. The tenacity of these issues and of the organizations that contest them is often disproportionate to their importance, and seem rather to be organizational battles for prerogatives rather than concern for the common good.

4. SOME ORGANIZATIONAL AND INSTITUTIONAL CONSIDERATIONS IN TRANSPORTATION[14]

In this section we highlight some important organizational and institutional considerations specifically germane to transportation.

- The division of labor among federal, state and local transportation organizations needs to be re-examined and likely changed. In many metropolitan areas, the state DOT runs the freeways (sometimes with operations tools like main-line detection devices, ramp metering, etc., and sometimes not) and a state or suburban county runs the intersecting arterials. It would seem that, at the least, some proactive, positive system for coordinating these separately-operated systems needs to evolve, in the interests of both effectiveness and efficiency.
- Also, the federal role needs study. With a single focus since 1916 on capital investment in pursuit of national goals, such as interstate commerce and defense, an operations role and funding mechanisms for operations vis-à-vis state and local agencies needs to be defined. Steps to achieve this rebalance are underway. The idea of making transportation operations a core mission of the FHWA is moving forward, with the administration having "announced a restructuring and reorganization, the depth and breadth of which have not been undertaken in several

---

[13] Powell, James, "Implementing Coordinated VMS/HAR Operations in the Gary-Chicago-Milwaukee Corridor", ITS America Annual Meeting, 2001.

[14] This section draws upon Sussman, Joseph M., David Hensing and Douglas Wiersig, "Transportation Operations and the Imperative for Institutional Change", Working Paper for ITE Annual Meeting, Irvine, CA, April 2000.

decades"[15]. Illustratively, *Public Roads*, an FHWA magazine, in its May/June 2001 issue, had three articles on operations-related issues.[16]

- In considering operations vs. construction, our public-sector transportation organizations have understandably developed a culture over many decades in which construction is at the professional pinnacle and operations takes the table scraps. From a historical viewpoint, going back to "getting the farmers out of the mud" and the extraordinary (and successful) investment in the Interstate over a 45-year period, the construction focus was appropriate. Now, it needs to change. To change, we need to attract the best transportation professionals to the operations arena. Incentives, visibility and opportunity for promotion will need to attend any institutional change that will give a higher profile to operations.

- Our transportation organizations, particularly in the highway area, have focused on the provision of physical infrastructure and have developed a set of stakeholders, consultants and contractors who support the endeavor of new construction. With an operations perspective, these institutional relationships will need to change. With information about the transportation system being central to real-time operations and control of that system, the role of private-sector independent service providers is already emerging. The electronic linkage between vehicle and infrastructure requires cooperation between the public and private sector in the provision of high technology so that investments in infrastructure and in in-vehicle equipment will be compatible. The role of the public sector as decision-maker and the private sector as the doer of their bidding is changing. Private-sector for-profit opportunities and public/private partnerships will need to be part of our transportation institutional structures and operations mission.

- Our transportation agencies tend to be primarily modally-oriented. For example, the U. S. DOT has very strong modal administrations (FHWA, FAA, FTA, etc.) and a significantly more modestly-scaled Office of Intermodalism. However, transportation services for both travelers and freight, and the operations that support these services, are increasingly necessarily intermodal to provide high-quality service at a reasonable price. To accommodate this, our transportation agencies must reflect this

---

[15] Johnson, Christine M., "Transportation Operations: A Core Mission of the FHWA", *ITE Journal*, December 1999.

[16] Paulson, S. Lawrence, "5-1-1: Traffic Help May Soon Be Three Digits Away", *Public Roads*, May/June 2001;
Baker, William, "The ITS Public Safety Program: Creating a Public Safety Coalition", *Public Roads*, May/June 2001;
Winn, Melissa A., "Handling the Worst Crash Ever in Virginia", *Public Roads*, May/June 2001.

intermodal perspective, and important changes within these organizations, as well as in the relationships among them, will be required.

- We need to develop regionally-scaled transportation operating organizations, which will require previously unlinked organizations working together, sharing responsibilities and data. Further, these organizations need to be collectively capable of receiving and effectively dispersing funds for operations.[17]

This suggests a natural question: in this regional context, what then is the role for the *metropolitan planning organizations* with regionally-scaled responsibilities? Can we turn to them for the region operations mission? As Steve Lockwood notes, "...most MPOs have no tradition of involvement in operations-oriented projects..." and after all, these organizations were established as planning organizations. Can they transcend a TIP perspective?

The answer is some can and some cannot. The Metropolitan Transportation Commission in the San Francisco Bay Area can lay claim to being an operating organization. Some other MPOs are unlikely to be able to transition. If that transition is to occur, organizational *incentives* for change and the *resources* to perform are clear pre-conditions.

5. WHAT WE NEED TO BE SUCCESSFUL IN AN OPERATIONS MISSION

Operations needs to be *3F/3I/3R*.

- **FUNDED:** **First and fundamentally, of vital importance is** *continuing reliable financial support for operations.*

  Funding for Operations: The following figure contrasts the costs and benefits over time for infrastructure and operations.

---

[17] Those interested in a deeper understanding of ways to think about *regionally-scaled* transportation operations are directed to Appendix D.

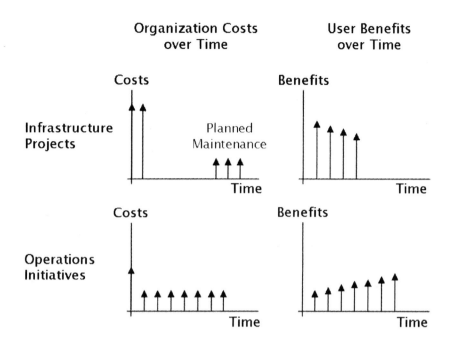

Conventional infrastructure projects have front-loaded costs for construction; then maintenance costs occur cyclically. Benefits accrue to users, but then may deteriorate between maintenance cycles and during maintenance procedures. The "coalition" for a conventional infrastructure project need be put together but once. A continuing stream of funds is not necessary, except for maintenance.

Operations funding is of a different character. Some up-front costs may be needed -- for example, some initial capital investment for ITS -- then there are year-to-year operations costs associated with transportation operations center staffing and the like. We suggest here that benefits associated with such initiatives may grow over time, in the sense that an operations focus allows modification to network processes as conditions change, while conventional infrastructure systems are unable to respond flexibly.

However, essential to this model is a continued flow of funds from year to year for operations. It is all too easy, as we have learned over the years, from deferred maintenance of conventional infrastructure, for operations funds to be cut in time of financial difficulty. It is quite difficult to extract value from an already-built piece of infrastructure and, conversely, quite simple to make a "temporary" cut in operations. If that occurs, the operations-based transportation services will quickly atrophy and customer service will deteriorate.

Boyden, et al.,[18] emphasize the need for a change in U.S. federal funding policy for operations. They note that we must do more than simply change organizations to deal with congestion on a regional geographic scale; we also need a change in the federal government, and specifically their funding mechanisms for highway operations. The authors note that the federal government does fund operations in transit, but in highways the funding mechanisms have been directed, for the most part, toward construction and maintenance.

- **FLEXIBLE: Operations are different every day. We need organizations that can respond flexibly to the many, diverse issues that arise. *Operations planning* is the way this flexibility is achieved.**

There are substantial differences between transportation planning and transportation operations. Most critically, they differ in time frame. Operations are ongoing; every day, operations begin anew. Planning deals with the strategic shaping of the transportation system. The product of planning is documents, while the product of operations is successful performance for another day's traffic.

However, within the planning framework, we should make an additional distinction between *strategic planning* and *operations planning*. Strategic planning deals with the development of plans that will guide the deployment of the transportation system over long periods of time, perhaps decades. Operations planning is planning for actually operating the transportation system. Operations planning guides the day-to-day operation of the transportation network.

> On June 26, 2001, *The Advocate* of Stamford, CT, printed a news story headlined, "Trapped Passengers Blast Metro-North", filled with complaints from passengers trapped in a Metro-North train for more than three hours. These passengers said the railway needed to "improve the way it responds to lengthy service delays".
>
> "'It was like this had never happened before and there wasn't even a plan,' said Wilton resident John Polich, who became stranded when the 10:53 a.m. train from Stamford to Grand Central Station snagged a wire about a mile east of Harrison, New York. 'The fact that they were behaving like this had never happened before was kind of shocking.'"

---

[18] Boyden, Edith, Allan DeBlasio, Eric Plosky, "Highway Funding: It's Time to Think Seriously about Operations", Volpe National Transportation Systems Center, Cambridge, MA, July 2001.

The article goes on to discuss the inability of the railroad to indicate how long the delay would last. Some crowd control problems arose.

Problems will happen with any complex transportation system, but *operations planning* -- designing mechanisms for responding quickly and effectively when those problems occur -- is fundamental. Mr. Polich's comment that "there wasn't even a plan", if correct, is a shortcoming of Metro-North. Operations means having contingency plans when things go wrong, and being able to respond in real-time with information and procedures to return to normal operations.

The parallel between transportation and manufacturing can be helpful to us in understanding how the operations mission changes organizational needs. Many authors, including seminally Arnoldo Hax and Harlan Meal[19], have written about the hierarchy of decisions in production planning, from design at a strategic level, to planning at a tactical level, and to operating decisions such as scheduling and dispatch. This notion of hierarchical production planning can be applied to the transportation field as well, where we have a comparable hierarchy of strategic network planning, tactical planning for operations, and transportation operations themselves.

TRANSPORTATION

STRATEGIC PLANNING
|
OPERATIONS
PLANNING
|
OPERATIONS
(E.G., ETC,
INCIDENT MANAGEMENT)

MANUFACTURING

PRODUCT DESIGN
|
PRODUCTION
PLANNING
|
OPERATIONS
(E.G., SCHEDULES,
DISPATCH)

PLANNING VS. OPERATIONS

---

[19] Hax, Arnoldo and Harlan Meal, "Hierarchical Integration of Production Planning and Scheduling", in M. Geisler (editor), *TIMS Studies in Management Science*, Volume 1, Logistics, New York, Elsevier, 1975.

- **FOCUSED:** **Discipline and focus is the required organization ethos for operations. This is a part of the 24/7 operations mindset.**

Operations Are Different: There are critical differences between operations planning and the act of *operations* itself. Organizations that have focused on strategic planning and operations planning may well have to undergo substantial organizational change to permit them to be effective *operators* of the transportation system. Operations are a fundamentally different set of activities and require a different organizational approach when compared with planning, construction or maintenance.

The Operations Mindset: A transportation system is a Complex, Large-Scale, Integrated, Open System -- a CLIOS. Operating it on a day-to-day basis is a difficult and complex undertaking. The system is subject to considerable uncertainty (e.g., weather), changes in demand, and variability in supply (e.g., because of road maintenance). Actions that one may take to ameliorate congestion in real-time may have counterproductive impacts on other parts of the system.

> CLIOS (Complex, Large-Scale, Integrated, Open Systems) is defined as follows[20]:
>
> A system is *complex* when it is composed of a group of related units (subsystems), for which the degree and nature of the relationships is imperfectly known. Its overall behavior is difficult to predict, even when subsystem behavior is readily predictable. Further, the time-scales of various subsystems may be very different (as we can see in transportation -- land-use changes, for example, vs. operating decisions).
>
> CLIOS have impacts that are *large* in magnitude, and often *long-lived* and of *large-scale* geographical extent.
>
> Subsystems within CLIOS are *integrated*, closely coupled through feedback loops.
>
> By *"open"* we mean that CLIOS explicitly include social, political and economic aspects.
>
> Often CLIOS are counterintuitive in their behavior. Often the performance measures for CLIOS are difficult to define and, perhaps, even difficult to agree about, depending upon your viewpoint. In CLIOS there is often human agency involved.

---

[20] Sussman, Joseph M., *Introduction to Transportation Systems*, Artech House Publishers, Boston and London, 2000.

Importantly, there is a *different mindset* associated with operations than the mindset associated with various planning functions. Since the latter is what many transportation organizations have traditionally done, we need to understand that distinction.

Simply put, the operations mindset is 24/7. It never stops. Although some difficulties are predictable (e.g., very heavy traffic the Wednesday evening before Thanksgiving), other events are not, and we must always be ready.

> The June 24, 2001 edition of *The New York Times*, in an article headed by "Repairs to Bridge on I-80 Will Mean Long Traffic Delays", discussed the damage to a small bridge on Interstate 80 caused by a major crash involving a gasoline tanker, and a subsequent fire. This dramatic event on a major roadway will doubtless have important and negative impacts on traffic in the New York Metropolitan area for weeks, if not months, to come. It is a critical operations issue.
>
> Yet all such problems are not of this major variety. In that same edition of *The New York Times*, an article on the opposite page was headed, "You Can't Get There from Here; Summer Brings Lots More Road Work". This article talked about much less dramatic, but nonetheless disruptive, *planned* operations in the New York Metropolitan area. A variety of streets are closed, as workers "dig, repair and repave". Maneuvering around these locally affected areas can be frustrating and time-consuming.
>
> Both these situations require an operations perspective, even though they differ in magnitude and nature. The first is concerned with an accident with damage to one structure on a major transportation link, Interstate 80, while the other refers to a great number of small, *planned* disruptions due to the need to repair street infrastructure. While different, they both require coordination among many organizations. This suggests the development of a plan for counteracting disruptions due to those link closures and the development of information that will allow effective operational planning to take place. So, for dramatic or mundane situations, we need an effective operations response and organizations that *can* respond.

The operations mindset involves what Matthew Edelman, Director of TRANSCOM, calls "blue collar regionalism", involving relationships between transportation operating organizations and police, fire, emergency vehicle organizations, and so forth. As Michael Ascher, head of the Triborough Bridge and Tunnel Authority, has said, "You are only as good as your last rush hour." Operations are every day. They happen in real time. They require decisive action and discipline to make the difficult operating problems routine.

> Consider the difference between a dermatologist and an emergency room physician. The dermatologist's patients "don't die, don't get better and never call in the middle of the night"[21]. The emergency room physician is 24/7, with life on the line every day. The professional mindsets of these two specialists are clearly fundamentally different.

- **INTEGRATED: An integrated organizational response is required to deal with operations in complex, geographically-diffuse transportation networks.**

The interconnections -- physical, informational and political -- in our regionally-scaled transportation enterprise determine their performance. Feedback dominates the mechanisms by which the transportation system creates value for customers. This suggests that operations for the transportation systems needs to be similarly integrated. This is a challenge. Organizations that have previously operated independently will now have to consider themselves as part of an integrated team, and perhaps additional feedback loops will need to be put in place. While difficult, this integration is a precondition for effective system performance.

- **INTERMODAL: Operations should be based on an intermodal concept. Our traveler and freight customers often view our services as intermodal, so we need to deal with transportation operations on this basis as well.**

For many years on the freight side, and more recently on the passenger side, the idea of creating superior customer service through effective intermodal operations has gained currency. Each mode has its distinctive advantages and disadvantages. The challenge of transportation operators is to put those modes together in such a way that advantages are maximized and disadvantages are minimized for each mode. An integrated trip chain is constructed. To achieve this, our operations focus must include an explicitly intermodal perspective.

---

[21] Thompson, Morton, *Not as a Stranger*, New York: Charles Scribner's Sons, 1954.

The Achilles heel of intermodalism has always been that the "hand-offs" between the modes are less than effective, thereby dissipating the inherent advantages of each mode. But new technologies -- and especially information and communications technologies -- allow these "hand-off" questions to be effectively addressed. The next section gives us our "third I" -- Information -- as a critical component of the operations focus.

- **INFORMATION- AND CUSTOMER-BASED: New technology has put into our hands the ability to collect, process and disseminate information to our *customers* and to the *partner agencies* concerned with operating the transportation network.**

This information provides an important opportunity to measure our performance using *customer-oriented metrics* and to greatly improve the quality of the service we provide.

Customer-Oriented Performance and Metrics: Our customers observe performance through the lens of operations.

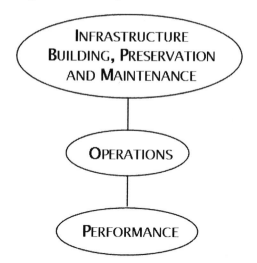

Planning and constructing the transportation network are *not* performance as seen by our customers. A Transportation Improvement Program (TIP) is a pre-condition to performance, but is not performance in and of itself. Yet many transportation organizations are geared to infrastructure planning and construction productivity metrics for measuring success. Changing that mindset requires organizational change to evolve to customer-based performance metrics, derived directly from the operations theatre.

Thinking through what specific performance metrics should be used is critical. It is well-known that what gets measured gets emphasized by the people within the organization and that, further, what does not get measured tends not to be worked on, *even if it is important.*

Some important characteristics of performance metrics are as follows:

- performance metrics should be designed such that managers can affect them. The manager needs to know *how* to change performance on that dimension.
- performance metrics should matter to the customer of the system, and further, should be something the manager is convinced matters to the customer.
- performance metrics should be easy and practical to measure.

Simply installing these performance metrics, no matter how cleverly drafted, will not be sufficient unless managers understand why these performance metrics are being implemented. Of course, through these performance metrics we are trying to induce managers to operate more effectively. In practice, it may be necessary to educate managers on how they, in an operational sense, can best respond. All the incentives in the world will not matter if we do not provide managers with some insight into how best to achieve better performance as measured by these metrics.

The new technologies available to us through ITS can be a great help in customer-oriented performance measurement. At the same time, it is important to *ask the customers* how they think the system is performing on various dimensions from time to time, and to ask them how *they* would measure performance from their perspective. Periodic surveys and focus groups are mechanisms for getting at the customer viewpoint.

- **REGIONAL: Operations need to be conducted at a regional scale, preferably using a regionally-scaled platform to support technical systems.**

Valerie Briggs (Kalhammer) describes how regionally-scaled organizations "note improvements in transportation emergency management and greater efficiency in operation", suggesting the criticality of that scale for operations management.[22] For further regional perspectives, the reader is invited to read Appendix D.

A Typology of Regional Operating Institutional Structure.[23] Here we present a typology of possible regional institutional structures. There are a variety of forms that regional operating institutional structures can take. We

---

[22] Briggs (Kalhammer), Valerie, "New Regional Transportation Organizations", *ITS Quarterly*, Fall 1999.

[23] This work is informed by Sussman, Joseph M. and Christopher Conklin, "Regional Architectures, Regional Strategic Transportation Planning and Organizational Strategies", ITS America Annual Meeting, Boston, MA, May 2000.

suggest here that there are four primary axes along which such structures may be characterized, as shown in the following figure.

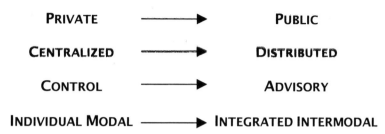

Any institutional structure can be represented on these four dimensions. First, it can be an integrated intermodal structure, including, for example, highway, public transportation, rail, and so forth. The antithesis of such a fully integrated structure are individual modal organizations with very loose or perhaps non-existent institutional connections. Of course, this is a continuum. Many institutional structures do not exist at either extreme, but rather somewhere between them.

The same continuum idea is applied to the other three dimensions. The structure can be a highly-centralized structure or it can be distributed, both physically and logically. The structure can be composed only of public agencies at one extreme, or only of private firms on the other, with a likely configuration in modern transportation systems being public-private partnerships. Finally, the structure can be based upon control of the transportation system (in the sense of command and control, as the military uses the term) or it can be simply advisory. In the latter case, the regional institutional structure provides advice to individual organizations within it, but they need not follow it.

Any institutional structure can be characterized by its position on the four axes. The institutional issues that we must deal with as well as issues within individual organizations are a function of the characteristics of a particular structure.

For example, one could characterize the New York Metropolitan area's TRANSCOM as
- Public
- Distributed (although virtually centralized)
- Advisory
- Modal (for the most part)
  Houston's TranStar would be characterized in this typology as
- Public
- Centralized
- Control (for the most part)
- Intermodal

- **REAL-TIME:** **Operating situations are ongoing, dynamic and driven by random factors. Real-time is the temporal scale on which operating organizations need to respond.**

Professional activity in transportation has long focused on strategic planning -- based on the time-scale of years or even decades -- and tactical planning -- recognizing that the Fourth of July fireworks or the ballgame may require special treatment of traffic on the next day. But now our operations focus requires a real-time perspective as our response to changing conditions must be very fast if congestion and safety and security hazards are to be ameliorated or avoided. Fortunately the technology exists, and more and more is in place to allow us to do real-time operations, but a change in professional training and in organizational and institutional perspective will be necessary for the real-time vision to be achieved.

- **ROUTINE:** **While individual stresses on the operating environment cannot be predicted, we can plan for generic kinds of service interruptions and treat them in a routine and decisive manner.**

We should minimize ad hoc-ism in our response to operating situations. Operations require planning. Knowing the chains of command and having contingency plans for "standard" situations is fundamental to transportation operations. If *performance* is to be achieved, the operations mindset, supported by effective *operations planning*, is a requirement.

> The $14 billion Central Artery/Tunnel project in Boston (the "Big Dig") is a major traffic challenge as its decade-long construction period proceeds. Keeping Boston running given the massive disruptions caused by the construction is a difficult undertaking.
>
> For five years the author chaired the "Traffic Mitigation" Committee of the Artery Business Committee, intended to be a watchdog on the Big Dig to minimize the effect of the construction process on traffic.
>
> Both the Central Artery/Tunnel project and the Boston Traffic Department have traffic operations centers (TOCs) intended to monitor traffic flows and to take appropriate action in the case of service interruptions due to incidents. Perhaps someday these will be coordinated, but currently they operate independently.
>
> One day several years ago, a contractor working on a section of the Artery south of Boston was slow in getting the

traffic barriers taken down after the night shift and it adversely affected northbound flows into the city during the morning rush hour period. Huge back-ups and delays were created.

As it happened, the day following this event, the Traffic Mitigation Committee met. The chief of the Central Artery/Tunnel traffic operations center and the chief of the Boston Traffic Department's traffic operations center both served on the committee. Both arrived at the meeting in high spirits. They proceeded to tell the story of how, through their joint efforts, they had allowed the city to recover from this major traffic disruption during rush hour. Without repeating the story in great detail, it involved these two senior officials running from site to site, speaking by cell phone in order to properly focus surveillance cameras to ascertain what was going on, getting senior police officials out of bed early in the morning to achieve a modicum of traffic control, and so on. These two senior officials were jubilant about the success they had achieved in limiting the delays caused by the bottleneck. But in thinking about traffic *operations*, to me, it represented a failure.

We must aspire to *routine* traffic operations. Disruptions should not require the first-in-commands of two TOCs to coordinate in an ad hoc fashion. The chain of command needs to be impeccably clear. People should know what to do. In short, *a routine disruption should be handled in a routine way*. This is achieved through *operations planning*. We should not need Butch Cassidy and the Sundance Kid, guns (or cell phones) blazing, to rectify the situation.

While this ad hoc collaboration between the Boston Traffic Department and Big Dig officials is somewhat understandable and likely temporary (given the Big Dig disruption), this is *not* a model for what we should aspire to in agencies charged with traffic operations.

6.   CHANGE IN FOCUS

But 3F/3I/3R does *not* say it all -- we need a fundamental change in focus to achieve our goals.

- **THE SYSTEMS APPROACH:** Operations is a *systems engineering concept* (as noted by Lockwood[24]). Our transportation organizations have not built strengths in this discipline. Initiatives by universities to provide this systems focus will be helpful (e.g., the focus on the systems approach in professional master's programs in transportation) but are long-term by their nature.[25] Continuing education of current transportation professionals is an important part of the focus change advocated here.
- **ACCOUNTABILITY:** Operating organizations must be *accountable* for their performance and need to self-assess to assure that performance. Organizational transparency is an important element of focus change.
- **INCENTIVES:** Organizational inertia and conservatism is a fact of life and it is sad but true that it is difficult to change the direction and mission of an organization simply because "it's the right thing to do". *Incentives* to lubricate change will be needed. Often, but not always, these incentives are financial in nature. *Federal financial incentives* are a vital part of focus change.
- **LEADERSHIP:** Visionary *leadership*, of the kind that gave us the organizations that produced the Interstate, is a key element of focus change -- and the success from earlier missions ironically makes the job of those change leaders all the more difficult.

7.   THE AGENDA[26]

The agenda for organizational and institutional change must be a strategic one, given the inherently slow pace of this change. Nonetheless, it is important that we embark on the course of evolving our transportation organizations and the relationships among them to create the *required* focus on operations. This agenda for change occurs at three levels: 1) the individual transportation professional; 2) the transportation organization; and 3) connections among these organizations and new organizations -- institutional change.

---

[24] Lockwood, Stephen, "The Institutional Challenge: An Aggressive View", Chapter 24 in *Intelligent Transportation Primer*, Institute of Transportation Engineers, Washington, DC, 2000.

[25] Sussman, Joseph M., "The New Master of Science in Transportation Degree Program at MIT", No. 1812, *Journal of the Transportation Research Board*, Washington, DC, 2002.

[26] This section draws upon Sussman, Joseph M., David Hensing and Douglas Wiersig, "Transportation Operations and the Imperative for Institutional Change", Working Paper for ITE Annual Meeting, Irvine, CA, April 2000.

1. **Professional Capacity Building.** We need to be explicitly concerned with the development of a new cadre of transportation professionals, focused on operations. These "New Transportation Professionals"[27], skilled in the technology and systems aspects of the transportation field, must have a sense of the institutional change required to achieve an operations focus in transportation organizations.

2. **Internal Changes to Transportation Organizations.** The magnitude of necessary change will vary among our current organizations. Some transportation agencies, at the state level especially, have had an historical focus on intercity issues, i.e., stitching the state and its neighbors together and leaving the issues within its metropolitan areas to others. Other agencies have emphasized building of facilities, with a corresponding de-emphasis on their operation. And still others have conducted transportation operations for years -- the latter largely at the metropolitan scale. The change required for a core operations mission in each of these kinds of organizations will be different in terms of time and resources needed, the challenge to leadership, and staff education and re-training needs. A fundamental mindset change is required. These changes will require strong leadership within these organizations, reflecting a substantial mission shift and a strong vision for the future of the role of that organization within the overall transportation system, as well as within our broader socio-economic-political system.

3. **Institutional Change.** While transportation organizations need to change their focus, with all the attendant difficulties in doing so, an even more difficult problem is redefining the relationships among the transportation organizations -- i.e., institutional change. Here, issues of control, budget, etc., rear their heads, but if we are to be truly effective in instilling an operations perspective, it is necessary that we create the interorganizational structures that can most effectively operate the transportation system, especially given our need to provide transportation services at a regional scale. These regional *partnerships are* new institutions and the change process is long-term.

Regional approaches have worked in some situations with respect to land use planning, infrastructure planning and even

---

[27] Sussman, Joseph M., "Educating the 'New Transportation Professional'", *ITS Quarterly*, ITS America, Washington, DC, Summer 1995. N.B. This article appears in Section III of this volume.

operations. We should learn from these successes. (See Appendix D.)

Change at the federal level will be necessary as well, as new funding mechanisms for operations are developed to provide financial support for the regional operations mission. These operating funds must be protected from cuts during budget downturns. This will require a new perspective on the part of FHWA and a changed perspective in the Congress on federal support for surface transportation; the latter should be embodied in the re-authorization of TEA-21. Federal leadership in creating incentives for regionally-scaled organizations to coalesce is an important step.

8.   A CLOSING WORD[28]

A focus on transportation operations is critical to the future of transportation. With limits on providing conventional infrastructure, an operations perspective that utilizes that existing infrastructure as effectively as possible is critical. Serendipitously, the technologies to allow us to do this, namely ITS, are now widely available. But the organizations that can take that operations perspective are, for the most part, still a gleam in our collective eye.

This operations perspective is required if transportation in the 21st century is to be customer-oriented, using performance metrics of relevance to travelers and shippers. This includes providing premium services to particular customers who are willing to pay a premium price for those services. The organizational and institutional implications of this change to a customer perspective are profound.

Moving toward the organizational and institutional change that will be needed is a long-term and difficult venture. It will require strong leadership and a vision for the future, in terms of how organizations are internally structured, how they relate institutionally to other organizations, and how they are funded. It will require substantial education and professional capacity building at various levels, and it will require viewing transportation at a regional scale. Indeed, Dr. Christine Johnson has suggested we need nothing less then a 3-C program for operations.

The changes we advocate are at three levels. Change at the individual level, where professional capacity building to develop "new transportation professionals" for our new organizations, will be essential. Our

---

[28] This section draws upon Sussman, Joseph M., David Hensing and Douglas Wiersig, "Transportation Operations and the Imperative for Institutional Change", Working Paper for ITE Annual Meeting, Irvine, CA, April 2000.

organizations must change internally to give greater emphasis to operations. Change needs to occur through new institutional connections to other organizations.

We call for a fundamental rethinking of our transportation organizations for the future -- at all levels of government and including the private sector -- asking them to participate in regionally-scaled transportation *operations*, utilizing new kinds of public-public and public-private partnerships and funding structures to create intermodal services for travelers and freight. This is no small thing to accomplish. Facilitating that kind of change through education and through strong and visionary leadership is central to the future success of the transportation field.

9.   AFTERWORD

Since the first draft of this article was written in July 2001, the world has surely changed in the wake of the tragic events of September 11, 2001. In the context of transportation, the urgency of that event emphasizes even more the need for an operations focus. The triplet of *security*, *reliability* and *safety*, which came out of the National Operations Summit in October 2001, *requires* an operations focus. Preparedness and system redundancy to provide secure, reliable and safe travel at a regional scale is all the more important.

Further, we have seen a continuing trend towards regional operations. Just one example is the article on AZTech in the *Newsletter of the ITS Cooperative Deployment Network*, entitled "AZTech Morphs into a Regional Operating Entity". [29] AZTech and others are showing the way in dealing effectively with the institutional and cultural requirements of operating transportation systems at a regional scale: recognizing the needs of customers, providing capacity without relying only on the building of conventional infrastructure, and introduction of new technologies. These regional organizations do not accept the status quo, as new partnerships are formed and institutional barriers are overcome. Leadership organizations are providing strong and positive examples of how to change our focus to transportation operations throughout the entire country. As noted above, education and strong and visionary leadership is critical if transportation is to continue to make its contributions to society in this suddenly more

---

[29] "AZTech Morphs into a Regional Operating Entity: A Discussion with Tom Buick and Dale Thompson", *Newsletter of the ITS Cooperative Deployment Network*, Fall 2001.

uncertain world. But there is ample evidence that our leadership is up to the task, and we are confident the barriers outlined in this document can be overcome through perseverance and a new vision for transportation operations.

APPENDIX A
LITERATURE REVIEW

The recent literature on institutional issues related to operations provides some useful insights.

Stephen Lockwood is a long-time and keen observer of the institutional scene in the transportation world. Some of his work over the last several years is particularly germane to the questions this article focuses on.

Lockwood in 1998 did an extensive report for AASHTO about state DOTs in the U.S.[30] In that report he notes the relatively slow rate of change of those state DOTs.

"In the United States, changes in the roles, organization, and processes of surface transportation institutions appear to be less dramatic (than in other countries). State departments of transportation (DOTs) are the dominant owners/operators of the principal surface transportation infrastructure. Their institutional context has been characterized by a long period of stability and a high degree of standardization, supported by a successful national program oriented to roadway system development and preservation have characterized their institutional context."

Lockwood further notes the challenge to changing these stable organizations. He suggests that among the toughest challenges are

" • Defining performance in customer-relevant fashion;
 ▪ Installing organizational incentives for change;
 ▪ Maintaining staff capability; and
 ▪ Institutionalizing mechanisms for continuous change."

He closes this report optimistically, suggesting that there is "the emergence of new models of organization, process, and relationships that reflect the special technical and institutional setting of surface transportation." Lockwood says that some of the key features in a changed state DOT would include:

" • Emphasis on real time operations of upper level systems using the best available ITS technology for reliability, safety, and security in conjunction with new multi-jurisdictional operating entities -- authorities or private corporations; ...

.
.
.

 ▪ Increased utilization of market mechanisms responding to customer willingness to pay (partnerships, tolls, commercialization) together

---

[30] Lockwood, Stephen, "The Changing State DOT", AASHTO, Washington, DC, October 1998.

with contemporary financial technology such as infrastructure banking, revolving funds and debt financing accessing nationally securitized capital markets;
- Incorporation of the best available technology in process activities (information systems), product development (material and process), and real time operation (intelligent systems); ..."

In a subsequent paper, Lockwood talks about ITS deployment[31]. Now, as noted earlier, while ITS is only a subset of an operations perspective, we argue much of what he says about issues in deploying ITS could be generalized to the operations context. He notes that an ITS program would be "substantially at odds with the current conventions of state and local transportation institutional arrangements with their major capital-improvement focus and related institutional arrangements and program structure". Lockwood calls for "a systems-engineering initiative, multi-stakeholder coordination and commitment to real-time operations" and suggests that such a change "cannot easily be accommodated within the institutional status quo". He goes on to discuss the challenges facing ITS and, by this author's inference, all of operations.

"These institutional challenges facing ITS can be described in terms of six 'preconditions' or factors that should be present, including:
1. An understanding of ITS concepts, elements and strategies, and the rationale for institutional change;
2. An authorizing environment formalizing the mission and providing the leadership, decision-making support and organizational structure;
3. New roles and relationships among various stakeholder agencies and entities necessary for effective ITS deployment and operations.
4. A planning and programming process adjusted to accommodate ITS-related strategies and investments competing for available resources;
5. Technology, staff and financial resources sufficient to support the deployment and operations of an ITS program; and
6. New public-private relationships as well as new private-sector business models responding to the specific potential of ITS.

There is no single institutional model through which these institutional challenges can be met. But there is no doubt that change is involved -- change that must begin with a vision and education and end in articulated, supportable programs."

There is much to be learned from reading Lockwood's work that is germane to this study of institutions and operations. He recognizes that substantial organizational as well as institutional change will need to take

---

[31] Lockwood, Stephen, "Realizing ITS: The Vision and the Challenge", *ITE Journal*, December 1999.

place for operations to be integrated into the mission of transportation organizations; that a systems engineering approach is fundamental to such a change; that there is no one right answer to what kinds of changes should be implemented; and financial resources will be essential to achieving any of these goals.

Lockwood has also written an extended chapter in the *Intelligent Transportation Primer*[32], which gives a great deal of detail about his perspectives on ITS and institutional change. For in-depth treatment of the issues noted above, this author suggests a careful reading of that chapter.

Gifford and Stalebrink[33] contribute to the transportation institutional literature by studying voluntary consortia and the concept of "learning organizations" considering the E-Z Pass consortia as an example.

Gifford, et al.[34], consider the issues that arise in a regionally-scaled traffic signal preemption and priority system in the Washington, DC, region. Preemption deals with allowing emergency vehicles and public transportation vehicles to preempt traffic signals, while priority allows for signal changes for specific vehicles (e.g., a bus which is behind schedule).

The authors trace the differences between the deployment in Virginia and Maryland, which have different institutional structures, and identify various system requirements needed to overcome institutional barriers to a deployment of these kinds of systems on a regional scale. These systems requirements include accountability (concern for unwarranted use by emergency personnel), interoperability (assurance that the systems can operate across jurisdictional boundaries), flexible and adjustable ("significant leeway in terms of its installation, its operations and the conditions for granting preemption"), ease of maintenance (concern that the system be "easily accessible and easily and inexpensively repaired or replaced"), clear control of operations and maintenance (simple coordination activities and responsibilities among the various traffic, transit and emergency agencies was viewed as critical). Further, "interference with the traffic community's ability to maintain and operate signals or require lengthy

---

[32] Lockwood, Stephen, "The Institutional Challenge: An Aggressive View", Chapter 24 in *Intelligent Transportation Primer*, Institute of Transportation Engineers, Washington, DC, 2000.

[33] Gifford, Jonathan and Odd Stalebrink, "Remaking Transportation Organizations for the 21st Century: Consortia and the Value of Organizational Learning", *Transportation Research A: Policy and Practice* (accepted for publication), 2001.

[34] Gifford, Jonathan, Danilo Pelletiere, John Collura and James Chang, "Stakeholder Requirements for Traffic Signal Preemption and Priority: Preliminary Results from the Washington, DC, Region", ITS America Annual Meeting, 2001.

coordination between agencies for routine maintenance" was anathema. Clearly the cooperating organizations wanted coordination among them to be as simple as possible.

Political issues arose in the deployment of this system. For example, some jurisdictions were concerned about granting priority to *underutilized* buses, causing problems with the general traffic flow. A critical issue was the interference of these preemption and priority systems with normal traffic light coordination.

Another useful paper was authored by *Wendell C. Lawther*[35]. Lawther emphasizes that there are many facets of operations, each of which may create different operating and institutional questions. He lists traffic light coordination, preemption/priority at traffic lights, freeway management, incident management, transit management, electronic toll collection, electronic fare payment, and regional multimodal traveler information, as components of an operating system.

Each of these operating tasks has different institutional issues associated with it. For example, preemption/priority of traffic lights, described above in the Gifford, et al., paper, will have different concerns associated with it as compared with electronic toll collection, where allocation of funds collected may be an issue .

Lawther emphasizes the need for *continuing funding for operations*, noting that these funds are quite different than capital funds. It is important to be successful in obtaining funds year-in and year-out to allow operations and maintenance to continue effectively, unlike capital funds, for which you need to develop a successful funding coalition only once.

Lawther also includes a useful and comprehensive discussion, along with good definitions, of various kinds of public-private partnerships.

Another paper of interest is by *James L. Powell*[36]. He examines variable message sign, highway advisory radio operations in the Gary-Chicago-Milwaukee corridor and considers the institutional questions concerned with deploying such a system in a three-state region with two major metropolitan areas. This system involved over 100 variable message signs and 12 separate highway advisory radio systems. The various stakeholders identified no less than *41* coordination issues. The paper goes on to describe

---

[35] Lawther, Wendell C., "Effective Public-Private Partnership Models in the Deployment of Metropolitan ITS", ITS America Annual Meeting, 2001.

[36] Powell, James, "Implementing Coordinated VMS/HAR Operations in the Gary-Chicago-Milwaukee Corridor", ITS America Annual Meeting, 2001.

those issues and how they were finally adjudicated among the stakeholders (in most cases).

In some ways Mr. Powell's paper is rather discouraging with respect to institutional issues. The 41 coordination issues seemed to this author to be mostly rather straightforward, dealing with uncomplicated technologies -- variable message signs and highway advisory radios. Yet the amount of coordination that was required, and the difficulty in reaching consensus was often daunting. This author was left with the feeling that if we have such difficulty in dealing with these seemingly straightforward issues, what hope is there for overcoming the deployment of more complex operations technology requiring much more autonomy-sharing and coordination?

While the shift to an operations mission and the deployment of ITS technologies implies important organizational and institutional change, we should recognize this is not the first time that the transportation field has faced such dynamism. A recent article by *Richard F. Weingroff*[37] traces the development of the American Association of State Highway Officials (AASHO) beginning in 1914 (now the American Association of State Highway and Transportation Officials (AASHTO)) and the development of the Federal Aid Highway Program several years later. Issues between the federal government and state government in those days included 1) the allocation of funds to individual states and an equitable formula for that allocation; and 2) the designation of "experimental post roads" as a mechanism to permit flow of federal funds to the states for highway purposes. Indeed, the federal Office of Public Roads and Rural Engineering (OPRRE) worked directly with many states to "draft state highway bills based on the OPRRE model state highway bill". It was noted that "OPRRE, which had been organized mainly for research and dissemination of information, would have to adapt to its major new responsibilities, which included the establishment of an engineering branch and a management and economics branch", changes comparable to those needed today.

So with the signing by President Woodrow Wilson of the Federal Road Act of 1916, an important partnership between the federal and state governments was formed, overcoming many organizational and institutional barriers. "As *Southern Good Roads* magazine said at the time, 'It will strengthen relations between the states and the nation, making them active partners in a great work for the common good.'" That partnership has endured over the decades. Certainly we can overcome some of the organiza-

---

[37] Weingroff, Richard F., "For the Common Good: The 85th Anniversary of a Historic Partnership", *Public Roads*, March/April 2001.

tional and institutional issues associated with our mission changes as well, here early in the 21st century.

Valerie Briggs (Kalhammer) has written several articles that inform our thinking on operations at a regional scale. In the Fall 1999 issue of *ITS Quarterly*, Briggs outlined a number of factors that are important in considering new regional organizations, including the partnership structure, the management structure, the role of state DOTs, the role of metropolitan planning organizations (MPOs) and the regional framework (architecture).[38] She suggests that regions that have developed regionally-scaled organizations "note improvements in transportation emergency management and greater efficiency in operation". She suggests that these organizations are "largely the result of local activity by authorities that operate transportation systems", rather then top-down imposition of such structures by the federal government.

Briggs (Kalhammer) has also written a short ITE paper, entitled "Operations in a Regional Transportation Organization Environment"[39], which emphasizes that institutional relationships are needed not only among transportation organizations (although these are important); an institutional design must also consider fire operations, emergency vehicles such as ambulances, law enforcement agencies, HAZMAT activities and towing companies. She emphasizes that real-time coordination and communication among the various organizations are important elements of operations. Briggs also discusses various organizational designs, which include physically centralizing the participating organizations under one roof, as is the case in Houston's TranStar, vs. the New York metropolitan area's TRANSCOM, which she characterizes as a "distributed, virtual transportation management approach".

In another more recent paper, entitled "Regional Operating Organizations: An Executive Guide"[40], Briggs (Kalhammer) defines three organizational types, including the virtual organization, the private corporation and regional government/authority, and describes the advantages and disadvantages of each. She develops case studies of TRANSCOM in the tri-state New York-New Jersey-Connecticut area, TransLink in

---

[38] Briggs (Kalhammer), Valerie, "New Regional Transportation Organizations", *ITS Quarterly*, ITS America, Washington, DC, Fall 1999.

[39] Briggs (Kalhammer), Valerie, "Operations in a Regional Transportation Organization Environment", *ITE Journal*, January 2001.

[40] Booz-Allen Hamilton, "Regional Operating Organizations: An Executive Guide", April 2001.

Vancouver, British Columbia, the Metropolitan Transit Commission (MTC) in the San Francisco Bay Area, the ITS Priority Corridor in Southern California, TranStar in Houston, and AZTech in Phoenix. She emphasizes the need for *integration* of resources, personnel, systems and institutions in order to achieve both user and institutional benefits. Briggs suggests that critical regional needs must be identified to develop support for regional operations. Such needs include congestion mitigation, environmental concerns, emergency response, and program implementation. She suggests that the inability to act upon those needs because of a lack of regionally-scaled organizations, will motivate the formation of such organizations.

Briggs also identifies the need for *funding*, especially for ongoing operations. She describes the decisionmaking structure of the various case study regions and focuses on the need for systems integration to deploy an effective operating system. She concludes with the statement that there are "multiple effective approaches to regional operations. The best approach will vary for each region depending on the transportation needs, available resources, and existing policies, procedures and institutional relationships of partners within the region."

John Wolf of CALTRANS has written an insightful and even inspiring paper about how to develop a transportation operations perspective within existing transportation organizations[41]. He makes a strong case for the operations perspective, calling it a "new world order", based on the *need for performance* of our transportation system. He discusses the importance of performance measure as a precondition for effective operations; his performance measures include mobility/accessibility, reliability, cost-effectiveness, sustainability, environmental quality, safety and security, equity, customer satisfaction, and economic well-being. Among the lessons Wolf identifies from his experiences at CALTRANS are the importance of data availability, integration across jurisdictions with local/regional efforts, ability to be truly modally-blind (that is, an intermodal/multimodal approach), thinking of the customer as a co-manager of the system, and internalizing all externalities (if one is to achieve livable communities and sustainability).

Lawrence Dahms and Lisa Klein, of the San Francisco Bay Area's Metropolitan Transportation Commission, describe the MTC's approach to a

---

[41] Wolf, John, "Performance Measurement and Integrated Transportation Management Systems: A Traffic Operations Perspective", CALTRANS.

regionally-scaled system for managing transportation in the Bay Area[42]. The MTC has long been viewed as in the leadership in development of a regional approach.   They emphasize new institutional structures to meet new management needs and the overarching criticality of *dedicated* funding for operations.   This theme persists in the papers about many regional organizations -- *the requirement that money be available to achieve the operations focus at a regional scale*. Operations monies are often difficult to obtain in the public sector, where the norm is capital expenditures for infrastructure.

This author notes that a reason for the success of MTC may result from the fact that it has had long-term continuing strong leadership in the person of Lawrence Dahms, who, until his recent retirement, had served at the head of MTC for 20 years. This author is reminded of the well-known case of the design of the public transportation system in Curitiba, Brazil, viewed as a model system, and one in which land-use planning and public transportation systems are closely integrated. Jaime Lerner, in various political capacities in Curitiba, served continuing for several decades, providing ongoing leadership, which allowed the deployment of these innovative systems.

Allan DeBlasio, in his contribution to "What Have We Learned About ITS?"[43] indicates some of the institutional challenges involved in deploying ITS and in developing an operations focus. The issues he emphasizes are:
- Awareness and perception of ITS
- Long-term operations and maintenance
- Regional deployment
- Human resources
- Multi-organizational relationship
- Ownership and use of resources
- Procurement
- Intellectual property
- Privacy
- Liability

These issues arose in the analyses done of the model deployment initiative programs, various non-technical barrier studies, and interviews with ITS deployment project managers.   As noted earlier, ITS is not equivalent to operations. However, many of these institutional barriers, and

---

[42] Dahms, Lawrence and Lisa Klein, "The San Francisco Bay Area's Approach to System Management", *ITE Journal*, September 1999.

[43] DeBlasio, Allan J., "What Have We Learned About Cross-Cutting Institutional Issues?", Chapter 8 in "What Have We Learned About Intelligent Transportation Systems (ITS)?", U. S. Department of Transportation, Washington, DC, 2000.

especially regional deployment, human resources and multi-organizational relationships, ring true in the operations context.

DeBlasio emphasizes that these barriers can be overcome and gives examples of situations in which they have been. He also emphasizes the long-term nature of these barriers and suggests "…that they will always be present in one form or another." He further emphasizes that the institutional changes required include new relationships between the public and private sectors and not simply public-public partnerships. DeBlasio cites especially the often difficult relationships between large public bureaucracies and small entrepreneurial firms that operate with different missions and different clock speeds.

DeBlasio further suggests the need for various organizations to cede some autonomy in regional coalitions and emphasizes the difficulties that often exist in accomplishing that.

In 2000 a workshop chaired by *Steven Gayle* was convened by the FHWA to consider linkages between planning and operations. Participants included state DOTs, MPOs, municipalities, transit authorities, and the FTA, as well as AASHTO, APTA, AMPO, ITE and PTI.

The group began by defining system management as follows[44]:

> " Effective transportation system management *maximizes transportation system performance* through a *coordinated and integrated decisionmaking approach* to 1) construction, 2) preservation, 3) maintenance, and 4) operation of transportation facilities, with the goal of safe, reliable, predictable, and user-friendly transportation."

The workshop participants noted that they had considered operations as subsumed within the overarching definition of system management.

The participants asked the question, "Why is system management not better integrated with transportation planning?" Various reasons were identified, including the following:

1. Transportation professionals have a project mindset -- or a "project culture" rather than a management culture.
2. "The constituencies for system management are very different than those for planning."
3. There is a fragmentation of operations responsibilities within transportation organizations. The workshop participants called this "fragmented ownership".

---

[44] "Summary of Workshop on Linking System Management/Operations with Transportation Planning", FHWA, 2000.

4. "The profession does not have adequate analysis tools to assess the effectiveness of system management strategies."
5. Often decisionmakers do not recognize the importance of system management, especially at the regional scale.
6. Transportation planning professionals have a variety of demands placed upon them: "environmental justice, air quality analysis, economic development, quality of life, etc." System management would just be one more thing "to be addressed by an already very busy planning staff".

Nonetheless, the workshop concluded that it was important that system management be a central function of transportation organizations, noting that a number of past initiatives (TOPICS, TSM, TDM, incident management, ITS) had taken place. System management focuses on performance and the public is demanding accountability for non-performing transportation systems. Various "big events", like the Olympic Games, show that we are able to perform the system management function when required to do so. The workshop urged a focus on *performance measures* for system management.

Finally, the workshop recommended pilots that could demonstrate the importance and impact of good system management. "The concept of market leaders focuses on those states and metropolitan areas that come close to the vision of integrated system management and planning as described above. Mini-case studies on national/regional symposia with representatives from these locations would highlight the message that linking system management and planning can be done and, most importantly, that it works!"

Literature on Organizations

There is an extensive literature on organizations. Understanding how organizations are *really* managed and how that may differ from that depicted by conventional organizational charts is the subject of research by *Henry Mintzberg and Ludo Van der Heyden*, documented in a *Harvard Business Review* paper[45], which suggests that the organizational chart is a poor way of understanding how companies work. They comment that "Indeed, using an org chart to 'view' a company is like using a list of municipal managers to find your way around the city... The fact is, organizational charts are the picture albums of our companies, but they tell us only that we are mesmerized with management. No wonder they have become so irrelevant in today's world." The authors define "organigraphs" as new ways of

---

[45] Mintzberg, Henry and Ludo Van der Heyden, "Organigraphs: Drawing How Companies Really Work", *Harvard Business Review*, Cambridge, MA, Sept-Oct 1999.

understanding how companies really work; this author suggests that this concept can be used in public-sector organizations as well.

They define the concepts of sets, chains, hubs and webs, and show examples of a petrochemical company, a newspaper company and a bank to illustrate how these concepts can be used to better understand the information flows and decision structures in organizations.

As we think about the new operations mission facing state departments of transportation, MPOs and other transportation organizations, it behooves us to integrate modern management thinking, such as that illustrated in this paper, into our organizational designs.

Other management literatures worthy of consideration in this context are:

1. Agency theory dealing with risk and responsibility sharing when the parties (e.g., organizations) have different attitudes toward risk and different goals.[46]
2. Technology-related strategic planning.[47]

---

[46] Eisenhardt, K. M, "Agency Theory: An Assessment and Review", *Academy of Management Review*, Vol. 14, 1989, pp. 57-74.
Reichelstein, S., "Constructing Incentive Schemes for Government Contracts: An Application of Agency Theory", *The Accounting Review*, Vol. 67, 1992, pp. 214-237.
[47] Christensen, Clayton M., Fernando F. Suárez and James M. Utterback, "Strategies for Survival in Fast Changing Industries", *Management Science*, Vol. 44, No. 12, Part 2 of 2, December 1998.

APPENDIX B

COMPANION PAPERS

As part of the same initiative that led to the writing of this paper, a number of other authors were also commissioned to write related papers about operations from various viewpoints. This section summarizes those papers, specifically from the perspective of the institutional issues identified by those authors.

"Highway Funding: It's Time to Think Seriously about Operations", Edith Boyden, Allan DeBlasio, Eric Plosky, Volpe National Transportation Systems Center, Cambridge, MA, July 2001.

In this paper, the authors emphasize the need for a change in U.S. federal funding policy for operations. They note that we must do more than simply change organizations to deal with congestion on a regional geographic scale; we also need a change in the federal government, and specifically their funding mechanisms for highway operations.

The authors note that the federal government does fund operations in transit, but in highways the funding mechanisms have been directed, for the most part, toward construction and maintenance. There have been earlier operations efforts, such as the TOPICS program.

The authors suggest some institutional issues, including:
- Defining operations;
- Developing institutions for operations to counterbalance the construction constituencies;
- The definition of new kinds of partnerships.

"Description of Transportation Systems Operations and Management", Steve Lockwood.

In this paper Steve Lockwood identifies the *fragmentation* of operations within many transportation organizations as an important institutional issue. He focuses on the need to integrate those operations, and the further need to develop performance measures focused on customers. Further, Lockwood discusses the importance of information-sharing among various organizations and the requirement for real-time response in the operations theater.

"Measuring System Performance: The Key to Establishing Operations as a Core Agency Mission", Michael D. Meyer, Georgia Institute of Technology, Atlanta, GA.

Michael Meyer focuses on major policy issues associated with the use of performance measures. He notes the importance of a shift toward a customer perspective; the implications of that shift include the need for accountability, efficiency, effectiveness, communications, clarity and continuous improvement over time. He cites various examples from around the country, including Minnesota, California, Florida, and Albany, New York. Meyer emphasizes system reliability as a relatively new performance measure in the highway transportation area.

"Data, Guns, and Money: A New Partnership for Highway Operations", Kevin Dopart, Mitretek Systems, Inc., Washington, DC, June 2001.
   In this paper Kevin Dopart considers the institutional barriers preventing the formation of a partnership between those organizations interested in *transportation operations* and those interested in *public safety*. He suggests that the public safety community *does not* see mobility as a priority, and discusses the lack of often even rudimentary communications among public safety and transportation organizations. Dopart argues that there are major advantages in more effective cooperation between these organizations. He suggests the need for incentives, possibly FHWA-provided incentives, to create an effective and leveraged partnership such that public-safety organizations recognize the desirability of mobility benefits in addition to their own direct responsibility for public safety.

"A Summary of Transportation Operations Data Issues", Richard Schuman, PBS&J, Winter Park, FL, July 2001.
   In this paper Richard Schuman emphasizes the linkage between better data and better operating decisions. To create exceptional performance, good data is essential, he argues. He notes the prevalence of "data gaps", by which he means coverage in both breadth and depth of the transportation network, and the lack of high-quality data available for decisionmaking. He argues for the need to "establish the environment that encourages innovation in data collection, sharing and use". He suggests the possibility of the public sector funding an initial "backbone" for ultimately public-/private-sector partnerships.
   He also focuses on data-sharing, particularly between the public and private sectors, and public and private roles in this process.

- "• *Data Sharing*. One of the key issues to a data collection program is the sharing of data between the public and private sectors. Who has the data? Who needs the data? What kinds of policies are in place to maintain the integrity of the data amidst the data sharing? Is there reluctance on the part of the public sector to share with the private sector? Under what limitations will the public sector allow the private to use the regional data? What are the constraints in sharing data between public sector agencies or the private sector sharing their data with the public sector? With the opportunity for multiple uses of data by multiple public and private organizations, data sharing has emerged as a major issue."

"• *Public/Private Roles.* Until recently, much of the network status data that has been discussed has largely been generated by public sector organizations using public resources. New data collection techniques, such as cellular phone probes or floating car data will likely be owned and operated by private sector data providers. Also, a federally sponsored program is facilitating the establishment of privately owned and operated roadside data collection infrastructure. How will the mixture of public and private data collection systems evolve? Can we create an environment that leverages public and private investment to improve the breadth, depth and quality of data?"[48]

"Traffic Congestion and Travel Reliability: How Bad Is the Situation and What Is Being Done about It?", Timothy Lomax, Shawn Turner, Mark Hallenbeck, Catherine Boon, Richard Margiotta, July 2001.

This paper, authored by Timothy Lomax, Shawn Turner, Mark Hallenbeck, Catherine Boon and Richard0 Margiotta, considers *why* congestion exists and tracks freeway performance as it has evolved over the last several decades. The authors note the condition of sprawl as a driver of congestion is, in and of itself, an institutional issue as local control over land-use leads to very spread-out land-use patterns in metropolitan areas. They suggest that traffic operations is a minor player in the sprawl game, with other factors being substantially more important. At the same time, they note that construction alone will not ameliorate the congestion problems that exist in our great urban metropolises

They consider operational and management strategies being pursued to relieve congestion, identifying the triplet of construction, improved operations and managing travel demand. Construction, of course, is difficult to rely upon, particularly in already-crowded urban areas. They cite operations strategies, including advanced traffic management systems, incident management systems, traveler information and managed lanes. They note that each of these can be helpful, but generally achieve only modest reductions in congestion.

---

[48] Schuman, Richard, "A Summary of Transportation Operations Data Issues", PBS&J, Winter Park, FL, July 2001, p. 14.

"Operations in the 21st Century -- Address by President at the 25th Annual Meeting of ITS America, July 4, 2025", Stephen Lockwood.

Finally, Stephen Lockwood, in the guise of a hypothetical speech by the President of ITS America in 2025, imaginatively describes the transportation world as it might be in 2025. He emphasizes the customer orientation of the 2025 system, together with "value pricing", used to make surface transportation more of a market-oriented system. He emphasizes the institutional changes that "have" taken place in that time interval, including "spanning the stovepipes" as "state and local governments took a variety of paths to aligning the operational elements of their existing programs". He also describes "institutionalizing new institutional roles", suggesting that

" The lure of federal transportation funds passthroughs stimulated new levels of collaboration among state and local governments including the law enforcement, fire and emergency services communities. These relationships rapidly progressed down the partnership spectrum from cooperation to consolidation. As they gained experience the concept of combined operations seemed less threatening and many smaller jurisdictions were happy to join forces as part of new regional operating entities. Even modal barriers were overcome, with transit and highway agencies sometimes taking on responsibilities for each other's services. Pressures to reduce government costs and size continued and outsourcing of facility operations and maintenance and TOC manning became standard practice in many areas. These arrangements provided the basis for important private sector investments and innovations." [49]

He goes on to say that financial incentives "were" helpful to achieve institutional change.

" Many institutional ideas, long discussed and dismissed, surprisingly gained momentum after demonstrations proved their benefits to skeptics (and blunted the objections of those vested in the status quo). Some of these included: the willingness of jurisdictions to consolidate operational management where substantial cost savings were available; the voluntary adoption of incident management practices by police and EMS agencies where reductions in autonomy were accompanied by significant subsidies from transportation." [50]

---

[49] Lockwood, Stephen, "Operations in the 21st Century -- Address by President at the 25th Annual Meeting of ITS America, July 4, 2025", p. 5.

[50] Lockwood, Stephen, "Operations in the 21st Century -- Address by President at the 25th Annual Meeting of ITS America, July 4, 2025", p. 6.

All in all, Lockwood presents an imaginative and plausible view of how institutions will change over the next quarter-century to provide an effective operations and customer-oriented surface transportation system.

Summary

To summarize the companion papers, the key points are as follows:
- To achieve effective operations, institutional change on the federal level, specifically with respect to funding for operations, is necessary.
- Operations are currently fragmented within transportation organizations. Consolidating operations will have an efficacious effect on services provided.
- Integration of transportation operations and public safety can lead to benefits for both activities.
- As a general proposition, there is a need for financial incentives to create institutional change, as in the one noted immediately above.
- The collection of high-quality data, so essential for good operations, presents institutional issues that need to be overcome -- again, financial incentives is a mechanism for doing so.

Many of the papers refer to the customer orientation that drives operations and the recognition of the importance of this in surface transportation as an enabler for the organizational and institutional changes needed to produce an operations-oriented surface transportation system.

APPENDIX C

DEFINITIONS

In this section we define some basic terms as they are used in this article.

Organizational change: By organizational change, we mean modifications to the structure and functionality of some organization from an *internal* perspective. For example, a change in management structure such that the "Chief of Operations" reports directly to the State Secretary in a state department of transportation, would be an example of an organizational change.

Institutional change: Here we refer to changes in the relationships among organizations. The design of new communications and operations relationships relating organizations in a particular metropolitan-based region would be included here.

Also included here would be the formation of new organizations that serve to integrate the functions of existing organizations. So the formation of TRANSCOM in the New York metropolitan area would represent an institutional change, as it was at the time, a new organization. Further, the new interconnections between TRANSCOM and, say, New York State DOT, would be another institutional change, but a new structure within the New York State DOT to accommodate that new interaction with TRANSCOM would be characterized as an organizational change for the New York State DOT.

Customer: A customer is any user of transportation services, including travelers and passengers, and organizations and people concerned with moving freight.

Stakeholder: Stakeholders include customers, but also encompass non-customers such as environmental interest groups, regional planning organizations, chambers of commerce, and so forth.

Operations: One of the dictionary definitions of operations is "to carry on operations in war; to give orders and accomplish military acts, as opposed to staff work". This definition seems consistent with our distinction between "planning" (a staff function) and "operations"(more of a line function).

Philip Tarnoff's and Dennis Christiansen's paper[51] defines transportation operations as follows: "Transportation operations can be defined as the application of techniques to facilitate the flow of vehicles, travelers and good on the existing surface transportation infrastructure". So here a distinction is

---

[51] Tarnoff, Philip and Dennis Christiansen, "Moving Toward a National Agenda of Transportation Operations and Mobility Research: Report of the Operations and Mobility Working Group", November 2000.

made between adding to the transportation infrastructure, as opposed to day-to-day operations of facilities.

Operations (an extended definition):   Operations of the surface transportation system covers a variety of functions.   In this extended definition, we consider *operating the infrastructure* -- the supply side of transportation; *managing demand* for transportation services; the management cycle of *measuring* performance, *reviewing* that performance, and *modifying* it appropriately; the communications functions with customer and stakeholders, and the *public relations and marketing* function. The list of activities is shown in the following table:

Operating Functions

- Operating the Infrastructure
    - Transportation Operations Centers (TOCs)
    - Electronic Toll Collection (ETC)
    - Network Management
        - e.g., Coordination of freeways and surface streets
    - Traffic Light Control
    - Emergency Response
    - Work Zone Management
    - Incident Management
        - Detection
        - Response
        - Removal
    - Intermodal Coordination

- Managing Demand
    - Road Pricing
    - Environmental Pricing
    - Traveler Information Systems

- Performance Measurement, Review and Modification

- Communication with Customers and Stakeholders, including Public Relations and Marketing Function

APPENDIX D

THE REGIONAL PERSPECTIVE

Perhaps the most difficult institutional issue we face is creating regionally-scaled transportation operations. In this appendix, we examine *ways of thinking* about the regional challenge. We consider the concept of *regional architectures* as a way of designing institutional structures that meet particular regional needs. Also, we suggest in the final sections, that while the emphasis of this article is operations, there can be other viable and complementary approaches to organizing regional transportation.

Regional Architectures: An Expanded Institutionally-Oriented View

In transportation, system architecture was originally a technical ITS-based concept to permit effective and technical interoperability of ITS deployments. (ITS adapted this concept from large military and aerospace contractors.) These system architectures were then used to create regional architectures, which also were technical in nature. Each region was asked to specify, through the architecture concept, the manner in which the technical elements of their system would interact, the required information flows, and the channels through which communications would take place.

But some authors have considered regional architecture to be more than a technical concept. They have defined it as an organizational and institutional concept as well. The idea has been used to characterize how *organizations* should interact within a regional framework to do planning and provide services. Jon Makler[52] defines the idea of a comprehensive regional architecture composed of a regional planning architecture (RPA) and a regional service architecture (RSA). The RPA and RSA define the ways in which organizations within a region interact for planning and service provision purposes, respectively.

---

[52] Makler, Jonathan, "Regional Architectures and Environmentally-Based Transportation Planning: An Institutional Analysis of Planning in the Mexico City Metropolitan Area", Master's Thesis, Massachusetts Institute of Technology, Cambridge, MA, June 2000.

COMPREHENSIVE REGIONAL ARCHITECTURE

REGIONAL SERVICE ARCHITECTURE   REGIONAL PLANNING ARCHITECTURE

from Makler, Jonathan
"Regional Architectures and Environmentally-Based
Transportation Planning: An Institutional Analysis of Planning
in the Mexico City Metropolitan Area", Master's Thesis,
Massachusetts Institute of Technology, Cambridge, MA, June 2000.

The figure shows overlap between the regional service architecture and the regional planning architecture since some of the same organizations may participate in both planning and service provision. As Makler notes, "A state department of transportation, for example, is likely to be a very active participant in planning activities and is also likely to be an important manager of transportation services, making it part of both RSA and RPA. However, it is possible that the office or division of DOT that participates in the RSA is a different unit of DOT than that which participates in the RPA. When communication between the RSA and RPA is important, it may be incorrect to assume that the presence of one organization in both subsets provides a reliable link."

A detailed discussion of how either an RPA or an RSA would be developed is beyond the scope of this article, but in outline form the process has the following steps[53]:

Stage One: Identifying the Organizations

We generate an inventory of existing organizations with various responsibilities. Organizations from the public, private and non-governmental sectors should be included.

Stage Two: Characterizing the Organizations

Organizations can be characterized on a geographic scale, according to its accountability to the body politic, and its management philosophy.

---

[53] Sussman, Joseph M. and Jonathan Makler, "A Tale of Three Metropolitan-Based Regions: Their Mobility Philosophies and 'Regional Architectures'", ReS/SITE Working Paper, Massachusetts Institute of Technology, Cambridge, MA, 1999.

Stage Three: Characterizing the Linkages

Here the information and control flows among organizations need to be defined, as do decision-making hierarchies, as applicable. This step allows one to assess the *capacity* of the collective organizations in the region.

Stage Four: Prescribing New Institutional Needs

Looking collectively at these organizations allows one to consider whether a new organization need be formed, which would be a radical institutional change, or whether, by changing the mission of the individual organizations and changing the linkages among them, one can meet the goals of the region for providing effective and efficient transportation services. The typology described in the previous section can be helpful in this stage.

"The concept of regional architecture suggests that all of the [organizations] can be represented as nodes on a network and the relationships among them as links between the nodes. The product of the methodology is a set of prescriptions for the architecture that improves the capacity of the institutions or creates new institutions and linkages among them to facilitate [transportation operations]." [54]

Various Regional Approaches Oriented to Operations, Land-Use and Infrastructure[55]

Regions have taken different approaches to regional transportation management. TranStar is a regional organization based in Houston, Texas, that has innovated institutionally. This organization coordinates *operations-oriented mobility strategies* among a number of different public-sector transportation and emergency response organizations, including the City of Houston, Harris County, Texas Department of Transportation, and METRO (which is the metropolitan transit company). This organizational innovation is focused on *operation*, with a broad regional scope and rather indirect political accountability.

While TranStar's innovation in institutional relations focused on *operations*, the main focus of this article, it is useful to recognize that other regions have innovated institutionally on other dimensions.

For example, in Portland, Oregon, the organizational innovation is through METRO (Portland's MPO), which focuses on *land-use* planning at the scale of the Portland metro-based region; METRO has direct political

---

[54] Sussman, Joseph M. and Jonathan Makler, "A Tale of Three Metropolitan-Based Regions: Their Mobility Philosophies and 'Regional Architectures'", ReS/SITE Working Paper, Massachusetts Institute of Technology, Cambridge, MA, 1999.

[55] This section draws upon Sussman, Joseph M. and Jonathan Makler, "A Tale of Three Metropolitan-Based Regions: Their Mobility Philosophies and 'Regional Architectures'", ReS/SITE Working Paper, Massachusetts Institute of Technology, Cambridge, MA, 1999.

accountability. In Atlanta, through the Georgia Regional Transportation Authority (GRTA), the Atlanta metro-based region coordinates *infrastructure* development across the region, breaking the stranglehold that local municipalities and counties have traditionally had over transportation infrastructure development at a regional scale. GRTA is given veto power over certain local (county) planning decisions. Its political accountability is indirect in nature.

One can see that these organizational innovations differ on several dimensions. First, as noted above, they differ in terms of philosophy, with Houston focusing on operations, Portland on land-use, and Atlanta on infrastructure, as the foundation of their overarching regional strategy. Related to this, the timeframes within which those philosophies operate differ as well. Houston's timeframe, given its operations focus, is short; Portland's is quite long, with a 40-year land-use plan in place; and Atlanta, with its focus on infrastructure, can be characterized as medium-term.

Further, the organizations charged with implementing these philosophies have varying levels of direct accountability to the public. The following table summarizes the approaches of these regions.

Region	Organization	Philosophy	Time Frame	Accountability
Houston	TRANSTAR	Operations	Short	Very Indirect
Portland	METRO	Land Use	Long	Direct
Atlanta	GRTA	Infrastructure	Medium	Indirect

We recognize that these three philosophies can be complementary -- using ideas from all can be a useful regional approach. And, of course, in Houston, Portland and Atlanta, ideas from each are used. The intent here was to contrast the philosophical differences.

References

1. Baker, William, "The ITS Public Safety Program: Creating a Public Safety Coalition", *Public Roads*, May/June 2001.
2. Booz-Allen Hamilton, "Regional Operating Organizations: An Executive Guide", April 2001.
3. Boyden, Edith, Allan DeBlasio and Eric Plosky, "Highway Funding: It's Time to Think Seriously about Operations", Volpe National Transportation Systems Center, Cambridge, MA, July 2001.
4. Briggs (Kalhammer), Valerie, "New Regional Transportation Organizations", *ITS Quarterly*, ITS America, Washington, DC, Fall 1999.
5. Briggs (Kalhammer), Valerie, "Operations in a Regional Transportation Organization Environment", *ITE Journal*, January 2001.
6. Bunch, James A., "ITS and the Planning Process", Chapter 26 in *Intelligent Transportation Primer*, Institute of Transportation Engineers, Washington, DC, 2000.
7. The Centre for Sustainable Transportation. *A Definition and Vision of Sustainable Transportation*, July 2001.
   http://www.cstctd.org/CSTmissionstatement.htm.
8. Christensen, Clayton M., Fernando F. Suárez and James M. Utterback, "Strategies for Survival in Fast Changing Industries", *Management Science*, Vol. 44, No. 12, Part 2 of 2, December 1998.
9. Conklin, Christopher, "Regional Architectures, Regional Strategic Transportation Planning and Organizational Strategies", Master's Thesis, Massachusetts Institute of Technology, Cambridge, MA, August 1999.
10. Dahms, Lawrence and Lisa Klein, "The San Francisco Bay Area's Approach to System Management", *ITE Journal*, September 1999.
11. DeBlasio, Allan J., "What Have We Learned About Cross-Cutting Institutional Issues?", Chapter 8 in "What Have We Learned About Intelligent Transportation Systems (ITS)?", U. S. Department of Transportation, Washington, DC, 2000.
12. Dopart, Kevin, "Data, Guns, and Money: A New Partnership for Highway Operations", Mitretek Systems, Inc., Washington, DC, June 2001.
13. Eisenhardt, K. M, "Agency Theory: An Assessment and Review", *Academy of Management Review*, Vol. 14, 1989, pp. 57-74.
14. Fleischer, Peter B. and Robert Hicks, "Operations and Management: What Does It Mean for Local Agencies?", *Public Technology, Inc.*, FHWA-OP-00-028, 2000.
15. Gifford, Jonathan, Danilo Pelletiere, John Collura and James Chang, "Stakeholder Requirements for Traffic Signal Preemption and Priority:

Preliminary Results from the Washington, DC, Region", ITS America Annual Meeting, 2001.

16. Gifford, Jonathan and Odd Stalebrink, "Remaking Transportation Organizations for the 21st Century: Consortia and the Value of Organizational Learning", *Transportation Research A: Policy and Practice* (accepted for publication), 2001.

17. Gómez-Ibáñez, J. A. and J. R. Meyer, *Going Private: The International Experience with Transport Private Delivery*, The Brookings Institute, Washington, DC, 1993.

18. Hanshaw, Stephanie and Steven Shapiro, "Benefits of Privatizing Operations Centers: The VDOT Model", ITS America Annual Meeting, 2001.

19. Hax, Arnoldo and Harlan Meal, "Hierarchical Integration of Production Planning and Scheduling", in M. Geisler (editor), *TIMS Studies in Management Science*, Volume 1, Logistics, New York, Elsevier, 1975.

20. Johnson, Christine M., "Transportation Operations: A Core Mission of the FHWA", *ITE Journal*, December 1999.

21. Lawther, Wendell C., "Effective Public-Private Partnership Models in the Deployment of Metropolitan ITS", ITS America Annual Meeting, 2001.

22. Lockwood, Stephen, "The Changing State DOT", AASHTO, Washington, DC, October 1998.

23. Lockwood, Stephen, "Description of Transportation Systems Operations and Management".

24. Lockwood, Stephen, "The Institutional Challenge: An Aggressive View", Chapter 24 in *Intelligent Transportation Primer*, Institute of Transportation Engineers, Washington, DC, 2000.

25. Lockwood, Stephen, "Operations in the 21st Century -- Address by President at the 25th Annual Meeting of ITS America, July 4, 2025".

26. Lockwood, Stephen, "Realizing ITS: The Vision and the Challenge", *ITE Journal*, December 1999.

27. Lomax, Timothy, Shawn Turner, Mark Hallenbeck, Catherine Boon and Richard Margiotta, "Traffic Congestion and Travel Reliability: How Bad Is the Situation and What Is Being Done about It?", July 2001.

28. Makler, Jonathan, "Regional Architectures and Environmentally-Based Transportation Planning: An Institutional Analysis of Planning in the Mexico City Metropolitan Area", Master's Thesis, Massachusetts Institute of Technology, Cambridge, MA, June 2000.

29. Meyer, Michael D., "Measuring System Performance: The Key to Establishing Operations as a Core Agency Mission", Georgia Institute of Technology, Atlanta, GA.

30. Mintzberg, Henry and Ludo Van der Heyden, "Organigraphs: Drawing How Companies Really Work", *Harvard Business Review*, Cambridge, MA, Sept-Oct 1999.
31. NCHRP Synthesis 296, Transportation Research Board, National Research Council, National Academy Press, Washington, DC, 2001.
32. Olmstead, Todd, "The Effects of Freeway Management Systems and Motorist Assistance Patrols on the Frequency of Reported Motor Vehicle Crashes", Doctoral Thesis, Harvard University, Cambridge, MA, May 2000.
33. Orski, Kenneth, "Congestion Relief Should Become an Explicit Objective of the Federal Surface Transportation Program", *Innovation Briefs*, Vol. 12, No. 5, Sep/Oct 2001.
34. Paulson, S. Lawrence, "5-1-1: Traffic Help May Soon Be Three Digits Away", *Public Roads*, May/June 2001.
35. Powell, James, "Implementing Coordinated VMS/HAR Operations in the Gary-Chicago-Milwaukee Corridor", ITS America Annual Meeting, 2001.
36. Reichelstein, S., "Constructing Incentive Schemes for Government Contracts: An Application of Agency Theory", *The Accounting Review*, Vol. 67, 1992, pp. 214-237.
37. Schuman, Richard, "A Summary of Transportation Operations Data Issues", PBS&J, Winter Park, FL, July 2001.
38. "Summary of Workshop on Linking System Management/Operations with Transportation Planning", FHWA, 2000.
39. Sussman, Joseph M., "Educating the 'New Transportation Professional'", *ITS Quarterly*, ITS America, Washington, DC, Summer 1995.
40. Sussman, Joseph M., *Introduction to Transportation Systems*, Artech House Publishers, Boston and London, 2000.
41. Sussman, Joseph M., "ITS Deployment and the 'Competitive Region'", "Thoughts on ITS" Column, *ITS Quarterly*, ITS America, Washington, DC, Spring 1996.
42. Sussman, Joseph M., "The New Master of Science in Transportation Degree Program at MIT", No. 1812, *Journal of the Transportation Research Board*, Washington, DC, 2002.
43. Sussman, Joseph M., "Transitions in the World of Transportation: A Systems View", *Transportation Quarterly*, Vol. 56, No. 1, Winter 2002, Eno Transportation Foundation, Washington, DC, 2002.
44. Sussman, Joseph M., "What Have We Learned About Intelligent Transportation Systems (ITS)?", Chapters 1 and 9, Synthesis report for the series of papers, EDL #13316, http://www.itsdocs.fhwa.dot.gov/jpodocs/repts_te/@5901!.pdf, U. S. Department of Transportation, Washington, DC, 2000.

45. Sussman, Joseph M. and Christopher Conklin, "Regional Architectures, Regional Strategic Transportation Planning and Organizational Strategies", ITS America Annual Meeting, Boston, MA, May 2000.

46. Sussman, Joseph M. and Joseph Folk, "Unreliability in Railroad Network Operations", a chapter contributed to *Analytic Foundations of Engineering Problems: Case Studies in Systems Analysis*, Richard de Neufville and David Marks, eds., January 1973.

47. Sussman, Joseph M., David Hensing and Douglas Wiersig, "Transportation Operations and the Imperative for Institutional Change", Working Paper for ITE Annual Meeting, Irvine, CA, April 2000.

48. Sussman, Joseph M. and Jonathan Makler, "A Tale of Three Metropolitan-Based Regions: Their Mobility Philosophies and 'Regional Architectures'", ReS/SITE Working Paper, Massachusetts Institute of Technology, Cambridge, MA, 1999.

49. Tarnoff, Philip and Dennis Christiansen, "Moving Toward a National Agenda of Transportation Operations and Mobility Research: Report of the Operations and Mobility Working Group", November 2000.

50. Thompson, Morton, *Not as a Stranger*, New York: Charles Scribner's Sons, 1954.

51. Weingroff, Richard F., "For the Common Good: The 85th Anniversary of a Historic Partnership", *Public Roads*, March/April 2001.

52. Winn, Melissa A., "Handling the Worst Crash Ever in Virginia", *Public Roads*, May/June 2001.

53. Wolf, John, "Performance Measurement and Integrated Transportation Management Systems: A Traffic Operations Perspective", CALTRANS.

54. _____ , "AZTech Morphs into a Regional Operating Entity: A Discussion with Tom Buick and Dale Thompson", *Newsletter of the ITS Cooperative Deployment Network*, Fall 2001.

55. _____ , "Saving Lives, Time and Money Using Intelligent Transportation Systems: Opportunities and Action for Deployment", Prepared by ITS America, February 2000.

II. 2. DEPLOYING THE TRANSPORTATION/ INFORMATION INFRASTRUCTURE[1]

Worldwide development and deployment of Intelligent Transportation Systems (ITS) is accelerating. In parallel with this phenomenon is the onset of a large-scale communications infrastructure. Simultaneous development of these two concepts will have profound implications. The author explains how ITS and the National Information Infrastructure (NII) can become natural partners in the quest for improved productivity and quality of life for a nation's citizenry.

INTRODUCTION

Over the span of human history, technology and transportation have been closely intertwined. Going back centuries the development of the horse bridle had profound impact on geopolitical structures. More recently, the development of extraction technologies for oil, the rubber tire, and the internal combustion engine have all had critical impact on transportation as we know it today.

Today we see important technological advances that are in the process of shaping the modern transportation system. These advances include:

- The ability to sense system conditions on transportation networks.
- The ability to communicate, process and display large amounts of information cheaply and quickly over long distances.
- Algorithms for processing this information to improve systems performance.

Sensing the flow of vehicles on urban networks by ground-based technologies or by global positioning systems is already a factor in transportation system design. Communications technologies are advancing at an extraordinary rate. A recent article in *The Economist* speaks of "the death of distance", as the transmission of large amounts of data reliably, cheaply, rapidly, and over long distances becomes commonplace.

---

[1] Reprinted with permission of ITS America. Sussman, Joseph M., "Deploying the Transportation/Information Infrastructure", *ITS Quarterly*, ITS America, Washington, DC, Spring 1996.

The linkage of communications and information technology is changing the face of both these industries, as witness IBM's strategy of "Networkcentric Computing", (reported in *Business Week* last October 30) based on high-speed, low-cost digital networks.

Finally, advances in systems analysis and operations research have given us a means for using this information to improve transportation system operations.

EMERGING FACTOR OF INTERMODALISM

Technology is not the only changing factor in our dynamic world. Increasingly, our international economy has fueled major trade flows all over the world. The use of intermodal technologies to efficiently handle those flows is growing at a dramatic rate, and the idea of Intelligent Transportation Systems (ITS) as an enabling technology for such intermodalism is on the agenda. Emphasis on productivity in national economies as nations compete for markets and the thrust toward privatization of many services are all factors of concern. All have implications for ITS.

Currently, there is worldwide development and deployment of ITS. In parallel with this, many nations are establishing large-scale communications infrastructure. The thesis of this article is that these two critical developments in transportation and communications should not proceed independently. There are many advantages to coordination and joint development of these two important activities.

THE TRANSPORTATION/INFORMATION INFRASTRUCTURE

The parallel development of advanced transportation and communication/information systems provides some extraordinary opportunities worldwide. This combination was first called the *Transportation/ Information Infrastructure* in the *1992 Strategic Plan for IVHS in the U.S.*, prepared by IVHS (now ITS) America.

In analysis of the transportation/information infrastructure it is convenient to use U.S. terminology, namely, Intelligent Transportation Systems (ITS) and the National Information Infrastructure (NII). The former deals with the use of advanced information systems, communication and sensor technology to improve the effectiveness and efficiency of surface transportation. The latter deals with the provision of the ubiquitous set of communications capabilities that would in principle allow virtually universal interconnectedness.

Simultaneous development of these two concepts has profound implications for the development of each. Both ITS and NII can be critical components in the drive for national efficiency, the ability of companies to compete in the international economy and achieving improved productivity and quality of life of a nation's citizenry.

COMMON ISSUES AND INTERACTION

Some of the issues relating to these two concepts and the ways in which they may interact are as follows:

- Both require major investments for deployment and are characterized by the need for network-oriented development.
- Both depend on the ability to develop standards permitting interconnectedness and interoperability on a broad scale, possibly internationally.
- Both systems have heavy political content, given their substantial interaction with the population at large. Therefore, both have the (political) requirement to produce benefits broadly and provide universal basic service across the population.
- Protection of the public interest in both NII and ITS is necessary.
- The deployment of both is important to national productivity and in maintaining a nation's competitiveness in the world economy.
- Privacy and data security issues are important in the deployment of both systems.
- Substantial human resource development will be required to provide the technological support for both ITS and NII.
- The market for providers of both technologies is international in scale. Opportunities for buyers and sellers of ITS and NII exist worldwide.
- In both systems the development of new relationships between the public and private sectors will be necessary for effective deployment. Both have the requirement for participation by the public and private sectors for reasons of investment needs, regulation, and other institutional and political issues.
- Some suggest that ITS and NII can provide complementary services through the substitutability of information technology for transportation infrastructure, including concepts such as telecommuting and remote office sites.

- Both technologies will require jurisdictional cooperation, given the broadbased geographic areas over which both will be spread. Effective relationships among political jurisdictions must be established to assure proper interconnectedness and network efficiency. The requirements that NII and ITS provide broad-scale coverage in both urban and rural areas are important.
- Both systems are heavy spectrum users; careful management of this resource is required for effective deployment.
- Effective and equitable demand management systems must be developed for both ITS and NII. Despite the provision of high-capacity transportation and communication systems, history proves that the demand for such services, particularly at peak times, will outstrip the ability of the network to economically provide capacity.
- ITS architectural concepts can be extended to NII. The need for interoperability, interconnectedness and standards, as well as balanced local versus central control, has relevance in both ITS and NII.

A NATURAL PARTNERSHIP

From the above, it is obvious that ITS and NII comprise a natural partnership that can be developed with mutual benefits. The network structure and the ubiquitous character of both systems strongly suggest that cooperation by the system developers will be advantageous to both.

The worlds of transportation and communications are changing dramatically. Technologies such as cellular digital packet data (CDPD), asynchronous transfer mode (ATM), the Internet, and the concept of information-intensive systems dealing with the transmission of data as well as the transportation of goods and people, will shape the networks of the modern world. The common backbone of transportation and communication systems is an advantage that can enhance the development of both.

The relationship can go well beyond this, however. It is clear that communications needs for transportation facilities in an ITS environment are quite substantial. The need of a broad-band communications capacity for video and other data interchange is considerable. Some have suggested that ITS will be such a good customer of NII that it can have a good deal to say about the structure and pricing strategy of NII. ITS can provide the "content" that NII seeks.

Also, the transportation system is often an owner of right-of-way that can readily be used to provide communications infrastructure, such as fiber optic. Consequently, transportation authorities worldwide have recognized that they may well have a substantial role in NII deployment, while providing for their own communications needs.

Some transportation infrastructure organizations are considering building their own communications system to save the high costs of leasing such capacity from private-sector service providers. Some are considering entering the communications business as providers of communications services in addition to transportation services.

Still others are considering their role in facilitating public/private development, using their ownership of physical right-of-way and their own communications needs as mechanisms for building such partnerships.

NEED FOR INSTITUTIONAL FRAMEWORKS

There is the need for a new set of institutional frameworks to properly guide this parallel development of transportation and communications networks. Transportation authorities must consider a variety of issues in deciding which organizational structure is best. As noted by Melcher and Roos, these issues include capital costs and continuing costs, reliability, direct accountability to the public for the quality of services, operation and maintenance of both the transportation and communications infrastructure, resource availability, future upgrading of technology and additional need for capacity.

While the emphasis here has been on the parallels and commonalities involving ITS and NII, differences do exist. Transportation and communications industries often differ institutionally.

While differences clearly exist among countries, the surface passenger transportation in the United States -- largely highway and transit -- is typically a public system. The communications industry is often private and for-profit, although regulated. The rate of technological innovation in communications has been faster than transportation; the "half-life" of this technology is shorter than that of transportation technology, at least historically.

In forming new institutional relationships, these differences as well as the similarities must be considered. On balance, however, the combination of ITS and NII as a transportation/information infrastructure is a powerful idea.

THE EDUCATIONAL CHALLENGE

There is a need to develop highly-skilled professionals that can properly deploy communications and transportation systems. In my article

"Educating the 'New Transportation Professional'"[2], it was argued that a new generation of transportation professionals is needed to deal with the challenges of ITS in a communications-intensive age.

These professionals must understand the systems concepts that enable ITS deployment and the benefits and risks inherent in these advanced networks. They must also be adept in understanding the institutional structure within which the transportation/information infrastructures will be deployed.

This is a tremendous challenge to academia as well as public and private organizations. But just as there is a need to invest strategically in technologies, there is a need to invest strategically in human resources and use ITS and NII as opportunities to educate this new generation.

CONCLUSION

Transportation systems address the need for economic development, improving quality of life for a nation's citizenry, including the limiting of environmental impact. Modern transportation systems, being information-intensive, must be deployed within the framework of modern communication network technology.

The intersection between Intelligent Transportation Systems and the National Information Infrastructure is important. ITS as a major user of the NII, and transportation authorities as right-of-way owners provide grounds for cooperation between the ITS and NII. Judgments about ITS and NII should be made in a coordinated fashion. There are distinct advantages in doing so from the viewpoint of economic development and resource allocation.

New institutional relationships will be required as the previously separate transportation and communications fields begin to coordinate their needs. Effective interaction between the public and private sectors and various levels of government is also essential.

These institutional barriers may be difficult to overcome. However, the benefits of effective cooperation between these two major system developers make the effort worthwhile.

---

[2]   Sussman, Joseph M., "Educating the 'New Transportation Professional'", *ITS Quarterly*, ITS America, Washington, DC, Summer 1995. N.B. This article appears in Section III of this volume.

ACKNOWLEDGEMENTS

This article was informed by a conference at Harvard University in July 1995, sponsored by the Volpe National Transportation Systems Center, Harvard and MIT, and was adapted from a speech by the author at the Second ITS World Congress at Yokohama last November. The author also acknowledges useful discussions, advice, supporting papers and other information involving John Jensen, IBM and a Fellow at MIT's Center for Advanced Engineering Study; Dr. Thomas Mottl, Stratec Consulting; Dr. Thomas Horan, Claremont Graduate School; and Dr. Lee McKnight, Center for Technology, Policy and Industrial Development, MIT.

II. 3. ITS DEPLOYMENT AND THE "COMPETITIVE REGION"[1]

Intelligent Transportation Systems (ITS) is a mature international program. Programs in Japan and Western Europe date from the mid-1980s and the U.S. program received major impetus with the signing by President George H. W. Bush of the Intermodal Surface Transportation Act (ISTEA) in 1991. Certainly we have learned a great deal about ITS during this time period. But perhaps the most succinct lesson is that ITS research and development is much easier than ITS deployment. Certainly there remain unsolved research questions. But as one speaks to ITS experts around the world (including researchers), the recurring theme is

"Why is it so difficult to deploy ITS?"

We argue that the fundamental strength of ITS is also the most important barrier to its deployment. The fundamental insight of ITS is that the infrastructure and the vehicles that operate on it perform as a *system*. Technologies in the areas of computers, communications and sensors link together the previously independent vehicle and infrastructure components of surface transportation. Through that linkage, it is argued, substantial congestion reduction, safety improvements and productivity gains can accrue.

However, the systemic nature of ITS has other implications. Many have noted the distinction between ITS and the development of the Interstate program in the United States. The latter, a $130 billion highway development program, can be characterized as largely a public-sector initiative with the fundamental decisions being made by public officials. Conversely, ITS requires development on both the public-sector and private-sector sides, the former typically for infrastructure development and the latter for in-vehicle equipment. This coordinated rollout has proven difficult.

Resources are always an issue, but in ITS, resources in both the public and private sectors are required; scarcity of those resources in government units and private corporations is a constraint.

---

[1]   Reprinted with permission of ITS America. Sussman, Joseph M., "ITS Deployment and the 'Competitive Region'", *ITS Quarterly*, ITS America, Washington, DC, Spring 1996.

The systemic nature of ITS also requires new levels of sophistication in both public- and private-sector organizations. Many organizations have focused on either the infrastructure or the vehicle without much concern for how either would operate as part of an integrated system. To do so requires some organizational redesign; these kinds of institutional changes are difficult to achieve.

So the very systemic nature of this new concept, its fundamental strength, slows deployment.

Of course, there have been important successes in ITS. One can look at TRANSCOM in the congested New York/New Jersey/Connecticut region as a successful ITS deployment. The SmarTraveler program in Boston is an example of an advanced traveler information system with a strong initial track record. The Houston TRANSTAR is another example of an ITS deployment which is quantitatively and qualitatively improving transportation service in the Houston metropolitan area. Various deployments in Western Europe and Japan are advancing as well.

Nonetheless, it is fair to say that the transportation world is growing impatient with the relatively slow pace of deployment activity. It behooves us to think strategically about how deployments can be accelerated.

1.   THE CONTEXT FOR ITS DEPLOYMENTS

Let us consider how the world has changed since ISTEA in 1991 and how these changes might affect deployment strategies for ITS.
- We have seen a continued strong growth in international trade which has, in turn, accelerated the trend toward freight intermodalism as containers flow from truck to rail to ship around the globe.
- The development of what, in the U.S., is called the National Information Infrastructure (NII) is a worldwide phenomenon as countries around the world recognize the importance of an interlinked communications system on national and international scales.
- Particularly in the United States, the ideal of federalism and devolution are important political trends. The U.S. federal government is "pushing out" much power and funding responsibility for a variety of programs (including transportation) to smaller units in government.
- The management and economic literature has focused on the notion of "regions" as critical units of competition. Often, we speak of the "competitive region". The work of Professors Michael Porter and Rosabeth Moss Kanter at the Harvard Business School, emphasizes the idea that subnational units will compete economically on the basis of productivity and quality of life provided for its citizens.

2.   ITS AS A REGIONAL INITIATIVE

What might all these trends and ideas suggest for ITS and, in particular, ITS deployment?

I suggest we think of ITS as a regional initiative which would consider metropolitan areas plus their hinterlands (some have used the term "city-states" for this regional entity). This scale gets us beyond simply the questions of congestion in the core city and considers overall flows both of travelers and freight around a more extended geographic area. This perspective requires us to consider the interaction between freight and passenger mobility, which has sometimes been absent in ITS deployments. It also allows us to deal with a broader economic base as well as a larger set of private-sector enterprises in building a constituency for deploying ITS in support of the "competitive region".

This scale allows us to truly take advantage of the systemic nature of ITS -- linking vehicles and infrastructure. The "competitive region", with its newly-devolved powers, would focus on providing efficient and effective intermodal transportation services, integrated with an information technology/communication network -- what we have called the NII -- to provide high quality communication services -- another critical component needed to achieve a "competitive region".

ITS can serve as a technology integrator for regional transportation and communications services -- an ITS and NII synthesis, utilized to provide an intermodal freight and passenger system on a regional scale. Only those regions that create such systems of efficient transportation and communication can effectively compete in the 21st century.

3.   A CONCEPT OF REGIONALISM

ITS can be the catalyst for developing regional strategic plans that guide a long-term vision of integration of transportation and communications technology. The idea of a "regional architecture" derived from the national systems architecture can be the framework for assuring that technology growth to provide for both transportation and communications capabilities is done so that, in the long-term, such systems are nationally and intra-regionally interoperable.

ITS can build on the concept of regionalism as a fundamental unit of economic competitiveness. ITS can serve as an integrator of conventional infrastructure, communications needs and intermodalism fueled by international trade trends. ITS can provide the framework for regional strategic planning that provides for integrated technologies. The ITS community should take the lead in putting forward ITS as this framework for strategic planning; that perspective is our best hope for long-term deployment of ITS technologies.

The strategic vision for ITS is as the integrator of
transportation, communications and intermodalism
on a regional scale.

This is an ambitious vision; substantial leadership is needed to achieve the institutional and technological change essential for such an integrated result. Regional deployment achieved by system integration and institutional change should be the focus, with ITS serving as the linchpin of the "competitive region".

SECTION III. ITS: IMPLICATIONS FOR THE TRANSPORTATION PROFESSION AND TRANSPORTATION EDUCATION

"The man who graduates today and stops learning tomorrow is uneducated the day after."

-- Newton D. Baker

ITS can and already has had important effects upon the transportation profession. In this section, we consider the challenge in educating that "New Transportation Professional" in this expanded era of ITS. Then we go on to consider what kinds of transportation *faculty* we will need in our educational institutions to effectively provide that education, and what is the future of these transportation faculty as the transportation field evolves to the broader perspective of *engineering systems*.

For additional articles relating to these themes, also see Section IV-3, 8.

III. 1. EDUCATING THE "NEW TRANSPORTATION PROFESSIONAL"[1]

The world of transportation is witnessing exciting and dramatic changes. These changes suggest the need for many organizations to reinvent themselves. The author proposes that academia put its own house in order by attracting new colleagues into transportation education through research programs like those for Intelligent Transportation Systems (ITS). The "New Transportation Professional", he argues, will need a much broader and deeper education to participate effectively in transportation.

1.   INTRODUCTION

This is an exciting time for the transportation world, which is at the nexus of a number of critical issues facing the United States.

The tie between transportation efficiency and national productivity and the ability of the United States to respond to the international economic challenge it faces is increasingly strong. The end of the Cold War and the need to convert much of our technological capability to non-defense applications can have a profound impact on transportation system deployment over the next several decades. Further, transportation is seen as fundamental to many of the environmental and energy issues the nation faces. The accessibility of people from all economic and social strata to job opportunities is seen as an important equity issue in our nation.

At the same time that we in transportation face these new challenges, many new technologies and concepts are maturing to the point where they can have a profound effect on our ability to produce efficient and effective transportation. For example, many believe that the field of intelligent transportation systems (ITS) -- which applies modern communications, computer, and sensor technology to surface transportation -- can address many urban congestion and safety issues in highway and public transportation. New materials technology can have important effects on the railroad, highway, airline, maritime and transit industries, as can new propulsion systems and fuels. Important developments in the operations

---

[1] Reprinted with permission of ITS America. Sussman, Joseph M., "Educating the 'New Transportation Professional'", *ITS Quarterly*, ITS America, Washington, DC, Summer 1995.

research field, including new technologies in network analysis and risk assessment, can greatly improve our understanding of complex transportation systems.

But technology and analysis methods alone are not the answer. Institutional changes are needed if we are to properly address the challenges that face us. Many have spoken about a new set of relationships between the public and private sectors as innovative ways of deploying transportation services. New sets of relationships among various levels of government -- federal, state, and local -- will have to be explored as well.

The integration of new technologies and concepts into traditional transportation organizations will require fundamental changes in the missions of those organizations and substantial shifts in the kind of professional staff they will need. Indeed, some in academia see the need for a fundamental reinvention of transportation education aimed toward developing the transportation professional for the 21st century.

This "New Transportation Professional" will need a much broader and deeper education to participate effectively in the transportation field. Skills both in the new institutional realities of the transportation field as well as in the new applicable technologies and analysis tools will be required. In-depth knowledge will be required to address the increasingly complex transportation environment.

2.   THE SIX "I"S

The "New Transportation Professional" must deal with the world as shaped by the six "I"s.

First, this world is INTERNATIONAL, as trade and a changing geopolitical environment require us to think about transportation on a global scale.

Second, this world will be increasingly INTERMODAL, because to achieve effectiveness and efficiency in our transportation system we will need to build on the inherent advantages of various modes and focus on the intermodal connections between them.

Third, the INFORMATION/COMMUNICATIONS technological explosion is changing some of the fundamentals of transportation. In this era of distributed processing, fiber optic cable and the Internet, we need to rethink what we mean by a transportation network. Indeed, many have talked about communications as being the "fifth mode", to go along with highway, rail, air and water. Now, we speak of a transportation/information INFRASTRUCTURE, our fourth "I", which will link together the new technologies of information systems and communications with the conventional infrastructure technologies in our field.

Fifth, we deal in an increasingly complex INSTITUTIONAL framework with new kinds of relationships between and among the public and private sectors reshaping the way we deploy transportation systems.

Sixth, the world of transportation is driven by INNOVATION. Changes in technology and methods of analysis and operation drive important changes in transportation systems. We are required to innovate in order to ameliorate the negative environmental impact of transportation systems, while assuring that economic growth is not restrained.

3.   A FRAMEWORK FOR TRANSPORTATION EDUCATION

In this article, we focus on the role of undergraduate and graduate education in shaping the "New Transportation Professional".

As we begin, we should recognize what we do in transportation education cannot be divorced from the broader educational mandates that are shaping today's colleges and universities. I commend to you a recent book entitled *An American Imperative: Higher Expectations for Higher Education*, which captures a number of viewpoints about what higher education should provide for the U.S., including "creating a nation of learners", "values", broadly defined, and "technological literacy".

How do we go about designing a curriculum for the "New Transportation Professional"? First, history teaches us that we are unable to predict the details of the evolution of the transportation field. While we can reasonably assess long-term trends that will shape our industry, one necessarily does that in broad strokes. This implies that our programs focus on education in the broadest sense rather than training in a more narrow sense. As we educate students for advanced practice, we need to think strategically and not tactically about their long-term needs.

The implication of this is that we need to teach a framework for what I call "continuous learning". I distinguish here between the terms "continuous learning" and "continuing education". The latter is typically provided through short, specialized subjects that practicing professionals take to learn about new areas. This is an important component of the educational enterprise, particularly in these changing times.

Continuous learning, on the other hand, is defined as a self-motivated, self-study, individual action taken by practicing professionals in a more informal manner. For this continuous learning to take place, we have a responsibility to teach a fundamental framework for understanding the world of transportation to our students. In addition, we need to give them respect for a broad range of learning experiences and a broad perspective on transportation, bringing new concepts and philosophies, as appropriate, into the field.

What is this framework for understanding the transportation enterprise? We begin with *Transportation Fundamentals*. Understanding what the components of that system are (infrastructure, vehicles, power, fuels, control, etc.) and how they fit together is basic. This suggests understanding of the existing modes and the intermodal interfaces that connect them and the physical relationships and processes that drive them.

Further, we must teach the relationship of the transportation enterprise to markets (both freight and passenger) and level of service variables for different customers. Students need to be grounded in fundamental concepts such as capacity, congestion, network behavior, equilibrium, stochasticity, etc.

Transportation fundamentals includes the concept of transportation as a *complex system* and a framework for analysis of this system.

Then, building on these fundamentals, knowledge in the triad of *technology, systems and institutions* is needed, as shown below:

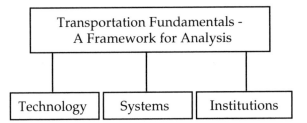

3.1. Understanding of technological advances and how they can be utilized within the transportation system

Technology has always been important. However, now its importance is multiplied several times over because
a. of the rate of change in technology
b. the pressures on the transportation system to work productively and in an environmentally-sound manner requires the integration of new technologies. Many of these are "high" technologies, a sophisticated set of advanced technologies that will fundamentally change our field and the kinds of knowledge that a transportation professional will need.

3.2. Understanding the systems methodologies fundamental to the analysis and design of transportation systems

This includes areas such as probability and statistics, optimization theory, microeconomics and various network analysis tools. Together with an understanding of these tools, we require a "modeling sense" which deals with the art of representing complex transportation systems in mathematical terms, both to gain insight into their performance as well as to work toward optimal modes of system operation.

3.3. Understanding of the broader context of where transportation fits in a societal/political/institutional framework

Here we focus on how transportation relates to its environment, broadly defined, and on providing a framework for understanding the process of rational policy development and how institutional change can be achieved. We need to recognize that institutional and policy analysis can be performed with rigor -- non-mathematical to be sure -- but rigorous nonetheless.

Developing these kinds of knowledge presents educators with a difficult task. Understanding the concepts identified above requires time and we are faced with the very pragmatic question of what to teach. There is in any degree program a "time budget" which limits the number of subjects and the research experience that students may undertake. That "time budget" constrains our ability to teach all of the above to the depth we would like. We have, as always, a "breadth-depth" trade-off in education.

How does one educate a broad transportation professional who is not superficial? How does one educate a specialist with in-depth competence in a particular area of transportation without creating a person with a narrow perspective on a field that is inherently broad?

Illustrative of the challenge facing us are the potential additional foci for transportation education programs beyond what many of us include now, such as:

- Modern communication systems and telecommunication policies
- Sensor and control technologies
- Advanced "engineered" materials and their uses
- Human factors
- Information networks
- Systems methodology extended beyond the current emphasis on planning and design to include real-time transportation operations
- Institutional issues in the context of advanced technology implementation

More topics could be added but this is already a daunting list; indeed, there is no magic answer as to what to include. Rather, each individual academic program seeks a balance appropriate to its institutional mission.

The same comment could be made about how one splits transportation education programs into undergraduate and graduate components. Different schools will design this split in different ways. We do note, however, that many current transportation graduate programs have flourished by enrolling students with a wide variety of undergraduate backgrounds, including engineering, science, social sciences such as economics and political science, and management. Many feel that there is great strength in this professional diversity.

The curse of the time-budget is somewhat alleviated by the concept of "continuous learning" described earlier. The "New Transportation Professional", properly educated, should have the tools and, if we have done our work right, the hunger, to learn in the future -- in a self-study mode -- about the complexities of the transportation world and related concepts. For example, some argue that a deep understanding of institutional issues comes only with professional maturity. Our graduate programs should prepare our students to develop these insights via "continuous learning".

The curse of the time-budget is further ameliorated by the educational role of the profession itself. The public and private transportation organizations which our students enter have some responsibility for the professional development of that individual and should provide an environment in which both continuous learning and continuing education are viable and encouraged.

4.   THE "T-SHAPED PROFESSIONAL"

To help think about transportation program structure, consider the concept of the "T-shaped" professional. (It is purely serendipitous that the letter "T" can be a useful construct for thinking about transportation educational programs.)

The vertical bar in the "T" represents an in-depth specialization of some subset of transportation. Creating that knowledge is important because it provides the intellectual rigor that will be needed to pursue innovations in our challenging field.

The horizontal portion of the "T" represents the breadth that we see as central to our educational process. The "New Transportation Professional" must have understanding at some level in transportation fundamentals -- technology, systems and institutions -- and must be sensitive to a broad set of issues that impact the world of transportation. Furthermore, that professional must have respect for the views of other professionals who have focused primary attention on different specializations.

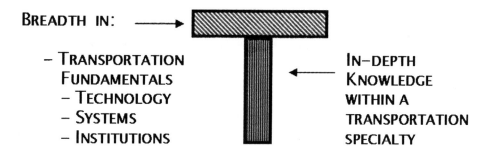

BREADTH IN:

- TRANSPORTATION
  FUNDAMENTALS
- TECHNOLOGY
- SYSTEMS
- INSTITUTIONS

IN-DEPTH
KNOWLEDGE
WITHIN A
TRANSPORTATION
SPECIALTY

5. THE ORGANIZATIONAL CHALLENGE TO ACADEMIA

We in education have a special challenge. We often advocate the concept of "organizational readiness" to our professional colleagues in the transportation field. We speak of the changing world of transportation and its new challenges and suggest that many of these organizations need to reinvent themselves to take advantage of the opportunities in this world. This is an appropriate role for academia, given our responsibilities for thinking strategically about the field of transportation. We are, in principle, well-positioned to make these suggestions to our colleagues in the profession.

However, we need to put our own house in order. Academic institutions, while good at advising change in other organizations, are often extraordinarily conservative considering changes to their own. Donald Kennedy, of Stanford, in his insightful article "Making Choices in the Research University" notes that the university will naturally gravitate toward strong disciplines in faculty appointments and away from more risky interdisciplinary choices. He characterizes the university as largely incremental in its approach to change.

> "The regular patterns of departmental choice suggest a powerful force in favor of incrementalism and against directed change. Collegiality is important in departmental function especially where -- as is often the case -- unanimous consent is a practical requirement for major decisions like new faculty appointments."

Kennedy goes on to note that the leadership in these research universities is not immune from this incrementalism in their view.

> "Even the best department chairs are more likely to be
> agents of their colleagues consensus than to be agents of
> change in some larger institutional context."

Therefore, we will need to be especially assertive in developing an interdisciplinary environment in which we can educate the "New Transportation Professional". I argue that to educate the "New Transportation Professional", we need to reinvent transportation education and the organizations within academia that provide that education. We need to re-emphasize the fundamentally interdisciplinary nature of our field and the roles of new technologies and concepts within it.

This suggests the critical need to attract faculty and research staff from other parts of the academic organization beyond that traditionally in the transportation field. Drawing those new colleagues into transportation through research programs like Intelligent Transportation Systems (ITS) is an important first step, but we must continue to integrate them within the broader framework of transportation education as well if we are to provide the educational program for initiating the career of the "New Transportation Professional". That approach has worked effectively in fields such as manufacturing and biomedical engineering, and ITS provides opportunity to make it happen in transportation as well.

This kind of education is "integrative", as described by President Charles Vest of MIT in his "Report of the President for the Academic Year 1992-1993".

> "(Our students must learn about) the analysis and
> management of large scale, complex systems. This means
> not only integration of the involved technical disciplines, but
> also an increased understanding of the larger economic,
> social and political and technical systems within which
> scientific principles and engineering analysis and synthesis
> operate in order to create technological change."

New organizational structures and new reward structures may well be necessary to achieve this. But we should not fear the change. The field of transportation is as vital and as front-and-center on the national and international stage as it has been for decades. This is a good time for this re-invention of transportation education.

The research university has a special role to play in developing educational programs for the "New Transportation Professional". Given its tying together of research and academic programs, these universities can and should play a leadership role in developing the inherently interdisciplinary education programs needed in transportation. However, we do this against the backdrop of a fundamental re-examination of the partnership between the federal government and the nation's major research universities. This is well underway, with some pain already being felt at many of these schools.

In the post-World War II era, Vannevar Bush wrote *Science: The Endless Frontier* as a framework for defining the research relationship between the federal government and universities that served the nation well for decades. It was focused on defense and essentially invented the idea of the research university in the United States. This government/university compact has run its course and redefinition is in order.

In the post-Cold War era, we are concerned with fighting an economic battle requiring a commitment to national productivity goals. Research on civil applications will be critical to winning this battle. This is the basis for a new compact between government and universities and including the private sector. Certainly, an extensive research program in transportation techno-logy, systems and institutions should be a central focus of this new compact. And a key output of such a research program will be new educational programs that educate the "New Transportation Professional" for the 21st century.

6.   WHAT IS DIFFERENT NOW?

So, as we approach the 21st century, what is different in the education of the "New Transportation Professional"? In the mid-1960s, transportation education evolved (in several schools) away from the detailed infrastructure design point of view to a more systems/economics perspective. That systems/economics perspective has largely shaped transportation graduate education for 30 years. Based upon operations research methodologies and economics, this framework has made a major contribution to advancing understanding of the complex domain of transportation.

However, many of the students educated in this way have overly focused on methodology and, in fact, confuse understanding of that methodology with the understanding of the fundamentals of the transportation field itself. We do not argue that these methodologies are not critically important, for, indeed, they are. We do argue that our students need to have a broader perspective on the field, including transportation fundamentals, transpor-tation technology and institutional issues.

Institutional issues are not new, just as technology is nothing new, but *the rates of change on these two dimensions is new* and these changes are fundamentally reshaping the field. The "New Transportation Professional" must have a firm grasp of these realities and opportunities.

Charles Handy of Oxford in his book *The Age of Unreason* talks about discontinuous change, as he notes:

"The changes are different this time: they are discontinuous and not part of a pattern. Such discontinuity happens from time to time in history although it is confusing and disturbing particularly to those in power...that continuous change requires discontinuous upside-down thinking to deal with it even if both thinkers and thoughts appear to be absurd at first sight."

These discontinuities are a characteristic of the transportation field and our education programs must reflect that.

7.   CONCLUSION

So, to conclude, our T-shaped "New Transportation Professional" must have the breadth to cover the areas described above as well as in-depth knowledge of some important aspect of the transportation enterprise.

An understanding of the world of transportation systems, together with analysis tools, a sound grasp of the institutional framework and knowledge in technology and its potential are the sine qua non of the "New Transportation Professional".

Changing our educational enterprise to properly capture this breadth of vision as well as in-depth understanding will require re-engineering of the transportation education organization. Many of us did this before in the 1960s and 1970s and we should not shrink from the challenge of doing it again.

The "New Transportation Professional" will be an educated person, as described in Peter Drucker's *Post-Capitalistic Society.* He or she will be a "knowledge worker" as Drucker defines it, providing value-added through a fundamental understanding of information and the need for it within the system under consideration -- in this case the transportation system.

I believe that the "New Transportation Professional" -- Drucker's "knowledge worker" -- should be educated as a *leader*, with in-depth knowledge, breadth, a taste for continuous learning, and vision as to the importance of transportation to modern society.

We close with a quote from Edmund Burke:

"A leader tells people not where they want to go, but where they ought to go."

The "New Transportation Professional" should be that leader.

References

1. *An American Imperative: Higher Expectations for Higher Education,* Wingspread Group on Higher Education, The Johnson Foundation, Inc., 1993.
2. *A Strategic Plan for IVHS in the United States,* ITS America, May 1992.
3. Bush, Vannevar, *Science: The Endless Frontier,* DC, National Science Foundation, 1945, Reprinted 1990.
4. *Creating a New Course in Transportation: Transportation Strategic Planning Services,* John A. Volpe National Transportation System Center, January 1993.
5. Drucker, Peter, *Post-Capitalistic Society,* HarperCollins Publishers, 1993.
6. Handy, Charles, *The Age of Unreason,* Harvard Business School Press, 1989.
7. The Urban Institute, *IVHS Staffing and Educational Needs -- Final Report,* Federal Highway Administration, September 1993.
8. Jovanis, Paul P., *Responding to IVHS Training Needs: A Curriculum for the 21st Century Professional Education,* IVHS America Annual Meeting, Atlanta, GA, April 1994.
9. Kennedy, D., "Making Choices in the Research University," DAEDALUS, Fall 1993.
10. *Re-Engineering Civil Engineering Education: Goals for the 21st Century.* Support Document for the 1995 Civil Engineering Educational Conference, September, 1994.
11. Smith, Brian L. and Lester A. Hoel, *Preparing the New Transportation Engineer: IVHS and Transportation Education,* IVHS America Annual Meeting, Atlanta, GA, April 1994.
12. Sussman, Joseph M., *The Future Challenges for Civil Engineering Education: Proposed Models,* ASCE Civil Engineering Education Conference, April 1985.
13. *Technology for Economic Growth: President's Progress Report,* Washington, DC, November 1993.
14. "The American Research University", DAEDALUS, Special Issue, Fall 1993.
15. Vest, C. M., *Report of the President,* MIT, Cambridge, MA, October, 1993.

III. 2. THE NEW TRANSPORTATION FACULTY: THE EVOLUTION TO ENGINEERING SYSTEMS[1]

In 1995, inspired by the changes in the transportation field induced by Intelligent Transportation Systems (ITS), Joseph M. Sussman wrote "Educating the 'New Transportation Professional'".[2] He described a broad concept of what knowledge the transportation professional in the 21st century will need to be effective. The article discussed the "T"-shaped professional; this professional would have a broad understanding of technology, systems and institutions in the transportation domain (represented by the horizontal bar on the "T") and in-depth knowledge in one of these specialties (represented by the vertical bar). With this combination of breadth and depth, the New Transportation Professional could contribute substantively to the solution of transportation problems, through detailed knowledge and an enhanced understanding of other professionals' contributions in what is inherently an interdisciplinary field.

Sussman concluded:

> "An understanding of the world of transportation systems, together with analysis tools, a sound grasp of the institutional framework and knowledge in technology and its potential, are the sine qua non of the 'New Transportation Professional'."

1. INTRODUCTION

The responsibility for educating the New Transportation Professional falls on the transportation faculty in colleges and universities. This article focuses on the "New Transportation Faculty" -- the educator of the New Transportation Professional.[3]

---

[1] Reprinted with permission of the Eno Foundation. Sussman, Joseph M., "The New Transportation Faculty: The Evolution to Engineering Systems", *Transportation Quarterly*, Eno Transportation Foundation, Washington, DC, Summer 1999.

[2] Sussman, Joseph M., "Educating the 'New Transportation Professional'", *ITS Quarterly*, ITS America, Washington, DC, Summer 1995. N.B. This article appears in this section (Section III) of this volume.

[3] In "Educating the 'New Transportation Professional'", Sussman argued that "continuous education" for the New Transportation Professional means that individuals will have the

What will that New Transportation Faculty need to *know* and *be* to carry out the educational mission? What is the individual instructor's future in academia?

To consider these questions, this article first traces recent developments in the transportation field. It also considers challenges in modern-day academia, and investigates the job of the transportation faculty as defined by the changing transportation and academic contexts. It contemplates the future the transportation faculty faces in a field that is mature, in a sense, fundamental, and yet ever-changing.

2.   TRANSPORTATION IN THE 21ST CENTURY

In the 1960s, many transportation faculty, recognizing the maturity of the physical design paradigm for transportation, led a radical change by introducing a *systems perspective* to both transportation research and education. This perspective focused on quantitative analysis of transportation as a complex, network-oriented system. Rather than stress the design of individual physical components, this approach recognized the systemic nature of transportation, and used operations research methodologies, along with transportation economics and simulation, as the appropriate tools for considering the planning and operation of the transportation system as a whole.   The approach was inherently integrative in nature, adopting methodologies from many areas. A great many students have been educated in this manner, giving rise to a number of educational programs around the country and the world.

But as the world moves forward, and the 21st century approaches, it is necessary to re-think this "new" system paradigm.

learning tools to evolve their professional skills independently over a career. Still, the basic framework will be put in place by the New Transportation Faculty.

3. TRANSPORTATION -- A BROADER AND DEEPER DEFINITION

Transportation has always been a vital force for the social, political and economic well-being of a society. It is a field long viewed as amenable to complex systems analyses that can give great insight into transportation investment, operations and design. However, the last 20 years have seen both a broadening and a deepening of the transportation field.

The transportation field now deals with operations on a global scale, considering the global economy and the implications for economic development and international competition among the regions and nations of the world. Indeed, one can argue that advances in transportation enabled this globalization trend. This geographic broadening has important implications for the future of the field.

Technology has always been a key element in the transportation field.[4] In the past several decades, advances in information technology and communications technologies have had profound implications on the way in which transportation systems operate. This will certainly continue. New opportunities for control of transportation systems -- particularly in real time -- is of fundamental importance.

Transportation has always had an impact on the fabric of society. In recent years, however, there has been a substantially increased focus on the *externalities* that relate to and shape transportation decisions. In addition to transportation impacts on economic development around the world, there is now heightened awareness of the importance of externalities in the areas of environment (with the importance of policies concerning clean air and water, and global warming), energy, societal equity, use of scarce land resources, and the fundamental tie of land-use to transportation. Society's concerns with all of these drive much of what is done in the transportation field.

Finally, there have been profound changes in the *organizational and institutional relationships* among entities concerned with providing and using transportation services. These changing relationships have brought about a re-thinking of public and private roles in transportation service provision, as embodied in such trends as deregulation and privatization of the transportation enterprise. From an intragovernmental point of view, there have been global trends towards devolution, as federal governments, recognizing the key role of regions in economic competitiveness, have

---

[4] Sussman, Joseph M., "ITS: A Short History and a Perspective on the Future", *Transportation Quarterly*, Vol. 50, No. 4, Anniversary Issue, 1996, ENO Transportation Foundation, Inc., Lansdowne, VA, p. 115-125. N.B. This article appears in Section I of this volume.

shifted power to regional entities, such as states, provinces, and large metropolitan areas.[5]

The trend towards intermodalism in both passenger and freight transportation, enabled by technology and demanded by the marketplace, requires important changes in interrelationships among transportation providers (e.g., rail, truck and ships in the case of freight intermodalism). Integrated supply-chain management, a central aspect of the logistics revolution, has fundamentally changed the relationship between transportation providers and their customers, for example, in the manufacturing industries, where transportation has become an integrated component of the overall logistics system of the customer.

All of these represent fundamental changes in the transportation field in *geographic* scale, *technological* scope, the *timeframes* for dealing with transportation issues, the relationships of the transportation system to *societal and economic priorities*, and the *organizational relationships* among transportation providers, customers and other stakeholders. These changes create new challenges for transportation researchers, educators and practitioners.

4.   EXPANDING THE BOUNDARIES

The above discussion suggests that a re-definition of the boundaries for the transportation field is needed. Transportation professionals and educators need to think on a broader, more ambitious scale on several dimensions. In addition, they need to create greater depth in models and frameworks for analysis, as the problem space becomes more complex and vital. They need to develop better methodologies for the increased scale of the transportation networks -- networks that must be controlled and operated in real time. Broader, more useful frameworks for qualitatively considering transportation issues need to be provided as well. Those in the transportation field need to focus not only on the operation of individual modes, but additionally on the way in which these modes interact intermodally, providing high-quality, low-cost service integrated with customers' operations.

A new definition of the field of transportation is needed. This definition must be both broader and deeper, more advanced technologically, and more sensitive institutionally. The view of the transportation field goes beyond

---

[5] Porter, Michael E., *Competitive Strategy: Techniques for Analyzing Industries and Competitors*, The Free Press, 1980.
Kanter, Rosabeth Moss, *World Class: Thriving Locally in the Global Economy*, Simon & Schuster, New York, 1995.

the systems analysis perspective invented in the 1960s, to a broader triplet in *technology, systems and institutions*. It addresses transportation issues

- at all timeframes, from real-time to strategic planning,
- on all geographic scales, from urban areas to the global,
- on all organizational scales, from the modal to the integrated supply chain,
- employing various approaches, including advanced models in operations research, simulation and econometrics, to consider transportation systems that are larger, more interconnected, and at the same time made more amenable to real-time control through new technologies in information systems and communications, and
- using qualitative frameworks of analysis to deal with the new and more complex institutional realities of the field.

As before, our educational approach is inherently integrative, but it is now on an even broader scale than before. A synthesis of quantitative and qualitative methodologies with transportation domain knowledge is required in order to deal in an integrated manner with the field.

The following figures graphically depict these approaches.

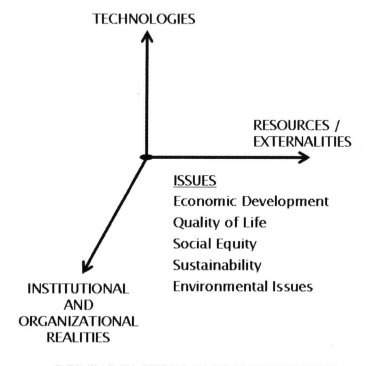

TECHNOLOGIES

RESOURCES /
EXTERNALITIES

ISSUES
Economic Development
Quality of Life
Social Equity
Sustainability
Environmental Issues

INSTITUTIONAL
AND
ORGANIZATIONAL
REALITIES

DRIVING FACTORS IN TRANSPORTATION

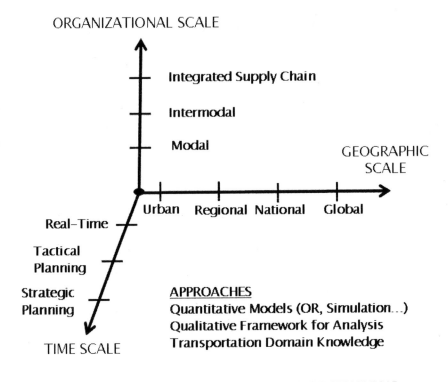

ORGANIZATIONAL SCALE

Integrated Supply Chain

Intermodal

Modal

GEOGRAPHIC
SCALE

Urban   Regional National   Global

Real-Time

Tactical
Planning

Strategic   APPROACHES
Planning   Quantitative Models (OR, Simulation...)
                   Qualitative Framework for Analysis
TIME SCALE   Transportation Domain Knowledge

TRANSPORTATION SYSTEM DIMENSIONS

5.   ACADEMIA

While the field of transportation has undergone remarkable changes over
the last three decades, these changes are no more remarkable than those in
the academic environment. The post-World War II period in the United
States gave rise to enormous growth in higher education and the
development of the American research university. Blueprinted by Vannevar
Bush during the war, and documented in his landmark paper "The Endless
Frontier", the research university established close working relationships
with the federal government, which actively supported research and graduate
education. Through this impetus, the research university became very
different from the classic university structure advanced by Cardinal John

Henry Newman, founder of the Catholic University in Dublin in the 19th century. A special section on universities in *The Economist*[6] notes:

> "Newman, like Cicero, believed in the need to separate the pursuit of truth from mankind's 'necessary cares'. His university would therefore be dedicated to the pursuit of knowledge for its own sake, would be 'the high protecting power of all knowledge and science, of fact and principle, of inquiry and discovery, of experiment and speculation'."

The Economist further explains, "…no institution will last nine centuries without adapting" and that "universities nowadays celebrate their achievements as producers of useful knowledge". Further, the university as a regional and national economic engine is a widely-accepted and successful model.

Many current-day senior transportation faculty became academics in the 1960s and 1970s, during the heyday of academic growth, fueled by federal support. Now, although universities are still "producers of useful knowledge", the close compact between universities and the federal government has atrophied as the Cold War ended with the implosion of the old Soviet Union.

As a result, the research university is reaching out much more actively to the private sector for the support of research in education. In addition, many have developed major programs abroad, reflecting the global economy and the place of the United States within that competitive environment.

This change in the support base implies the need for academia to be more relevant than ever and to respond more quickly than ever. Furthermore, U.S. society, feeling heavily invested in higher education, is looking to its universities for effective contributions to solve the many problems that contemporary humankind faces, often on a global scale.

The forces for change in academia are strong. Arthur Levine comments on the dramatic changes ahead for the academic profession.

> "Five forces are propelling the change: 1) the changing attitudes and demands of higher education patrons; 2) the changing characteristics of college students; 3) the changing conditions of employment in higher education; 4) the rise of new technologies; and 5) the growth of private sector competitors."[7]

Both government and the private sector have changed their perspectives on how relevant the universities should be and how quickly new competitors to universities are emerging.

---

[6] "Inside the Knowledge Factory", *The Economist*, October 4, 1997, pp. 3-5.
[7] Levine, Arthur, "How the Academic Profession is Changing", *Daedalus*, Fall 1997.

Theodore R. Mitchell characterizes higher education as "an industry under great stress" and notes:

> "Simultaneously, universities are challenged externally by the rise of competing providers of educational services, by increasing demands for short-term relevance in both teaching and research, and by the spread of new technologies that seem, at one level, to threaten the meaning of the university as a place."[8]

On another dimension, the technology of teaching has changed dramatically. A few years ago, it could reasonably be said that while a physician educated in 1890 would be lost in today's operating room, a professor in 1890 would find the classroom largely the same as it was in 1990. This is no longer the case. The use of the Internet and the WorldWide Web as educational instruments, and the growth of distance learning as professors routinely teach courses piped into a variety of industrial settings in real time, have forever changed the modalities of teaching. Some educators have even taught paperless subjects where assignments, readings and other materials, are all in electronic form.

In Charles H. Fine's recent book *Clock Speed*, he wonders, with the ability to project the image of world-class professors anywhere "...will our students still need to come to campus? For that matter, will the faculty still need to come to campus?" He further speculates whether or not, with world-wide fame, these faculty will even need a university affiliation![9]

Not everyone is convinced of the efficacy of this approach, and argue that the core competence of the university remains the residential college. While this may be so, changes in the way educators work are certainly in the air, and some argue that the modern university is in its death throes. No less a sage than Peter Drucker says, "Thirty years from now the big university campuses will be relics. Universities won't survive. It's as large a change as when we first got the printed book."[10] Drucker speaks of the growth in educational costs and comments, "...such totally uncontrollable expenditures without any visible improvement in either the content or quality of education means that the system is rapidly becoming untenable. Higher education is in deep crisis." He suggests, "The college won't survive as a residential institution."

The modern research university is recognizing its need to educate more effectively and also, given fundamental change in the nature of its financial

---

[8] Mitchell, Theodore R., "Border Crossings: Organizational Boundaries and Challenges to the American Professorate", *Daedalus*, Fall 1997.

[9] Fine, Charles H., *Clock Speed: Winning Industry Control in the Age of Temporary Advantage*, Perseus Books, October 1998.

[10] Lenzer, Robert and Stephen S. Johnson, "Seeing things as they really are", *Forbes*, March 10, 1997.

support, to be a central player in the solution of major issues facing society in a shorter timeframe.

It is in this search for university relevance that the importance of interdisciplinary, cross-cutting education and research becomes clear. It would be difficult to identify any critical societal issue that is not interdisciplinary at its base. Charles Vest, President of the Massachusetts Institute of Technology, states, "Humankind's advances will depend increasingly on new integrative approaches to complex systems, problems and structures. Design synthesis and synergy across traditional disciplinary boundaries will be essential elements of both research and education."[11]

6. TOWARDS THE NEW TRANSPORTATION FACULTY

It is in this challenge for academia that the "New Transportation Faculty" comes to the fore. Some argue that this kind of integrative approach is precisely what the transportation faculty has been about for the last 30 years, bringing together various technologies, mathematical approaches, microeconomics, political science, management and institutional studies into an integrated set of methodologies germane to the field.

The transportation field has become the exemplar of the nascent field of modern engineering systems, characterized by a cross-cutting, interdisciplinary approach to research and education. This integrative approach, motivated both intellectually and pragmatically as a result of changing relationships between academia, the federal government and the private sector, is the competitive edge for the New Transportation Faculty. Educators have shown their ability to integrate new knowledge into the field and to apply it effectively in their domain -- transportation.

Given this perspective, what is the future role for the New Transportation Faculty in academia? Think about what attracted many to the field of transportation. First, it was the notion of working on a real and serious societal issue in which improvements could have a direct effect on quality of life and economic development. Second, it was their interest in integrating new intellectual approaches, that is, new methodologies, into the study of that domain.

But there are broader challenges accessible to the New Transportation Faculty. Consider the idea of "Complex, Large-Scale, Integrated, Open Systems" (CLIOS). A system is *complex* when it is composed of a group of

---

[11] Vest, Charles, "MIT: The Path to Our Future", Report of the President, Academic Year 1997-1998, Massachusetts Institute of Technology, September 1998.

related units (subsystems), for which the degree and nature of the relationships are imperfectly known. Its overall behavior is difficult to predict, even when subsystem behavior is readily predictable. Further, the timescales of various subsystems may be very different (as in transportation, where, for example, land-use changes take place over years, while operating decisions are rapidly implemented).

CLIOS have impacts that are *large* in magnitude, and often *long-lived* and of *large-scale* geographical extent. Subsystems within CLIOS are *integrated* and closely coupled through feedback loops. By *open* we mean CLIOS include social, political and economic aspects.

Often CLIOS are counterintuitive in their behavior. At the least, developing models that will predict their performance can be very difficult to do. Often the performance measures for CLIOS are difficult to define and, perhaps, even difficult to agree on, depending upon one's viewpoint. In CLIOS there is often human agency involved.[12]

Transportation systems are simply a special case of CLIOS and the New Transportation Faculty have special capabilities to bring to the table. So as we look to the future, to go along with the T-shaped transportation professional, described earlier, consider the *I-Beam-Shaped New Transportation Faculty.*[13]

---

[12] Sussman, Joseph M., "ITS and Rescuing Prometheus", *ITS Quarterly*, ITS America, Washington, DC, Winter 1998. N.B. This article appears in Section IV of this volume.

[13] The "I-beam" image was suggested by Professor Lester Hoel of the University of Virginia.

Methodologies/
Domain Knowledge in Transportation

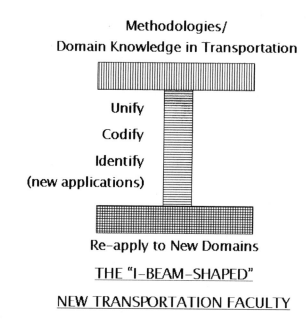

Unify

Codify

Identify
(new applications)

Re-apply to New Domains

THE "I-BEAM-SHAPED"

NEW TRANSPORTATION FACULTY

The top flange represents the disciplines that instructors have worked to integrate into the transportation domain. The web represents the unification and codification of these disciplines into a coherent approach to transportation applications and the identification of new areas within the engineering systems domain, to which the integrated approach can be relevant. Indeed, a recent report by the NSF calls for this codification.

> "Curriculum developments are needed that synthesize and codify knowledge from currently separate disciplines and fields, by leaders from those fields, to create appropriate basic or "core" subjects for transportation systems education. Another way to state this need is that the science base (broadly defined to include the social sciences) for the interdisciplinary field of transportation systems is in need of codification and presentation in a form suitable for graduate education. This development must be built on research developments of the kind discussed above. Subjects developed in this process will be useful both for transportation systems degree programs, and in departmental programs in transportation, and in some cases for continuing education."[14]

---

[14] "A Report to the National Science Foundation on the Workshop on Planning, Design, Management and Control of Transportation Systems", June 15, 1998.

The bottom flange represents a reapplication of these concepts to other applications' domains.

This is part of a broader reform in engineering education. William A. Wulf recently wrote eloquently of the need for incorporating a set of "new fundamentals" into the engineering curriculum.[15] This is certainly correct. It can be argued that these new fundamentals are the basis of a new field called "engineering systems", the integrated set of quantitative and qualitative methodologies, of which transportation systems is at the leading edge.

This integrated approach is valid in other applications. Transportation can be characterized by big infrastructure investments, a network structure for delivery of services, a (relatively recent) application of real-time control, a global scale, and a changing institutional structure, particularly between public and private sectors. Some or all of these characteristics are germane to other domains of societal relevance; telecommunications and energy are but two examples of systems with big infrastructure investments and a network structure. The urban region as a system could also be characterized in this way.

So the future for the New Transportation Faculty is clear. Building on our strengths in integrating new disciplines, faculty members can help create the new unified field of engineering systems and help in the application of this approach to a broad set of societal issues.

What will become of transportation if the New Transportation Faculty focuses on a broader applications domain? First, it must be understood that this is an *evolution*, not a *revolution*, and there is still much richness in integrating new ideas and codifying those ideas in the transportation domain. The well is by no means dry. Indeed, if one interprets the word "faculty" as a *collective* noun, there will always be individual faculty *members* focusing on the core transportation applications.

But more importantly, through the reapplication to broader issues of the engineering systems approach developed by transportation faculty, members will invent new concepts and discover new relevant disciplines that can be applied within the transportation domain. They will also attract specialists in those disciplines to transportation studies.

Those in academia should never lose sight of the fact that transportation is part of a broader set of societal systems. The interconnections between transportation and other domains are fundamental to the broad-based understanding of the field. There is no conflict in this evolutionary re-application of transportation concepts within a broader application space

---

[15] Wulf, William A., "The Urgency of Engineering Education Reform", *The Bridge*, Spring 1998.

and, indeed, still more interesting concepts can be integrated into the study of the transportation enterprise.

7. THE NEW TRANSPORTATION FACULTY'S TRADITIONAL ROLE

While there is a broader role for the New Transportation Faculty, there are traditional tasks as well -- *their job is still to integrate, innovate and instruct.*

To *integrate*, educators must constantly be alert for new methodologies and technologies that can be brought into transportation applications. They have done so effectively on the quantitative side of the field, bringing such methodologies as operations research, econometrics into the milieu. They have perhaps been less effective on the qualitative side in understanding the power of various kinds of political and institutional analyses as they seek to categorize transportation systems and the impediments to their development and effective deployment. Professor Tom Hughes, in his recent book *Rescuing Prometheus*,[16] brings a style of analysis for large-scale, technologically-based systems, including the Central Artery/Tunnel Project in Boston as an example of how qualitative analyses can advance the field.

As always, those in academia have a special responsibility to *innovate*. They push the envelope of the transportation domain and develop a deeper understanding of its behavior. Traditionally they do this through research, and it is an accepted principle that "through research, faculty members gain insight into the questions at the frontiers of their field, enabling them to build this excitement and focus into their teaching and coursework".[17] However, this innovation should extend to a new style of research concerned with advanced practice in the transportation field. Following from their colleagues in urban studies and architecture, transportation educators need to implement "studios" for our transportation students, working closely with faculty and practitioners on innovative transportation projects.

Finally, the transportation faculty has traditional role in *instruction*. With pressures to "solve society's problems" and to conduct cutting-edge research, educators must not lose sight of their fundamental mission; to educate students, with the classroom remaining an important component of that overall process. Transportation faculty members must diligently apply creativity to teaching in the same way they do to research. To quote from Donald Kennedy in *Academic Duty*:

---

[16] Hughes, Thomas P., *Rescuing Prometheus*, Pantheon Books, New York, 1998.
[17] "The Report of the MIT Task Force on Student Life and Learning", Massachusetts Institute of Technology, Cambridge, MA, 1998.

> "Great teachers exhibit, in their teaching, forms of creativity that may not be usually thought of as research, but nevertheless, they analyze, synthesize and present knowledge in new and effective ways."[18]

In the integrative world in which the New Transportation Faculty teach, such imagination in teaching is important and, indeed, it is helpful to the maturation and growth of all as faculty members. To again quote Kennedy, "Even in a world in which research seems to get the most attention, faculty members feel deeply rewarded if they sense they have made a difference in the lives of their students."

In addition to moving to become the I-Beam-Shaped New Transportation Faculty, educators must continue to *integrate* knowledge into the transportation domain, continue to *innovate* in research and in practice, and recognize the special importance that *instruction* has in this interdisciplinary field.

8.   CONCLUSION

The challenges of the transportation field continue and expand and the New Transportation Faculty -- the I-Beam-Shaped New Transportation Faculty -- has a special role therein in integrating knowledge, innovating, and effectively instructing the T-shaped New Transportation Professional.

But the challenges and opportunities are broader. Consider transportation as a special case of CLIOS and the special advantage the New Transportation Faculty has, given the inherently integrative nature of its intellectual approach and its track record in establishing how this integrated approach can impact the important societal domain of transportation. The future role of the New Transportation Faculty can include an engineering systems attack on societal issues using the CLIOS construct.

This positions faculty members to make important broader contributions to society and provides an intellectual growth path within academia, one that is in total concert with the new post-Cold War mission of the university for relevant, shorter timeframe approaches to important societal and industry problems.

---

[18] Kennedy, Donald, *Academic Duty*, Harvard University Press, Cambridge, MA, 1997.

Nor is the transportation field disadvantaged by this. The intellectual stimulation for the "New Transportation Faculty" can only make approaches to the core issues of transportation more effective.

References

1. Sussman, Joseph M., "Educating the 'New Transportation Professional'", *ITS Quarterly*, ITS America, Washington, DC, Summer 1995.
2. Sussman, Joseph M., "ITS: A Short History and a Perspective on the Future", *Transportation Quarterly*, Vol. 50, No. 4, Anniversary Issue, 1996, ENO Transportation Foundation, Inc., Lansdowne, VA, p. 115-125.
3. Porter, Michael E., *Competitive Strategy: Techniques for Analyzing Industries and Competitors*, The Free Press, 1980.
4. Kanter, Rosabeth Moss, *World Class: Thriving Locally in the Global Economy*, Simon & Schuster, New York, 1995.
5. "Inside the Knowledge Factory", *The Economist*, October 4, 1997, pp. 3-5.
6. Levine, Arthur, "How the Academic Profession is Changing", *Daedalus*, Fall 1997.
7. Mitchell, Theodore R., "Border Crossings: Organizational Boundaries and Challenges to the American Professorate", *Daedalus*, Fall 1997.
8. Fine, Charles H., *Clock Speed: Winning Industry Control in the Age of Temporary Advantage*, Perseus Books, October 1998.
9. Lenzer, Robert and Stephen S. Johnson, "Seeing things as they really are", *Forbes*, March 10, 1997.
10. Vest, Charles, "MIT: The Path to Our Future", Report of the President, Academic Year 1997-1998, Massachusetts Institute of Technology, Cambridge, MA, September 1998.
11. Sussman, Joseph M., "ITS and Rescuing Prometheus", *ITS Quarterly*, ITS America, Washington, DC, Winter 1998.
12. "A Report to the National Science Foundation on the Workshop on Planning, Design, Management and Control of Transportation Systems", June 15, 1998.
13. Wulf, William A., "The Urgency of Engineering Education Reform", *The Bridge*, Spring 1998.
14. Hughes, Thomas P., *Rescuing Prometheus*, Pantheon Books, New York, 1998.
15. "The Report of the MIT Task Force on Student Life and Learning", Massachusetts Institute of Technology, Cambridge, MA, 1998.
16. Kennedy, Donald, *Academic Duty*, Harvard University Press, Cambridge, MA, 1997.

SECTION IV. "THOUGHTS ON ITS", *ITS QUARTERLY*: FIVE YEARS OF COLUMNS ON ITS ISSUES

> "Highly complex engineering systems are very new, far from optimal and heavily constrained by both historical and nontechnical considerations."
>
> -- J. M. Carlson and John Doyle
> *Complexity and Robustness*

Section IV is different from the others in this volume. Rather than being built around a particular ITS-related theme, it contains 14 short columns in a series called "Thoughts on ITS" on various ITS issues that were important (in the eyes of the author) at the time of their writing (beginning in 1996). I believe all of these are relevant today as well. These columns often touch on ideas mentioned above, such as organizational issues, education, regional architecture, and try, in a succinct and often -- it is hoped -- pointed way, to capture the essence of the debates in these areas. Other columns deal with questions not touched on in a major way in this book -- issues such as safety, automated highway systems, where ITS fits in the science and technology strategy of the federal government, the ITS political frame, ITS as a complex, adaptive system, and ITS' role in developing countries. We hope these "probes" will give the reader a useful perspective on some important ITS issues.

IV. 1.   BEYOND TECHNOLOGY -- LOCAL ORGANIZATIONAL READINESS FOR ITS DEPLOYMENT[1]

At a number of recent ITS meetings there has been substantial emphasis on the central and critical role of local public-sector transportation organizations in the deployment of Intelligent Transportation Systems (ITS). This is certainly appropriate.   These organizations understand the local environment and needs for service as no one else can. As clearly articulated by Dr. Christine Johnson, Director of the U.S. Department of Transportation's Joint Program Office (JPO), putting ITS deployment in local hands is clearly the best strategy.[2]

For these organizations to be successful in ITS deployment they must change broadly, in terms of both core competencies and in perspective. In this connection, there has been a good deal of emphasis on "capacity building" at the local level.   What is usually meant by this is the development of technical ITS-related skills to allow local organizations to effectively implement advanced transportation systems.

There is no question that capacity building should be a priority. Ed Rowe, in his *IVHS Review* article three years ago clearly identified the lack of technical expertise on the local level as a key impediment to ITS deployment.[3]   Traditional skills in conventional infrastructure design, construction and maintenance must be augmented in areas such as information systems and communications.   The ITS community, and specifically the JPO and ITS America, through its Subcommittee on Educational and Training, and the academic world, has responded to these needs with short courses, expanded curricula, and other devices.   Good progress is being made.

However, we must go beyond expanding technical skills at the local level for ITS to be effectively deployed. I have used the term *"organizational readiness"* to describe the set of capabilities and organizational viewpoints

---

[1]   Reprinted with permission of ITS America. Sussman, Joseph M., "Beyond Technology -- Local Organizational Readiness for ITS Deployment", "Thoughts on ITS" Column, *ITS Quarterly*, ITS America, Washington, DC, Summer 1996.

[2]   Johnson, Christine, "Accelerating ITS Deployment: A Report from the U.S. DOT," *ITE Journal*, Dec. 1995.

[3]   Rowe, Edwin, "Operations and Maintenance: The Achilles Heel of ITS," *IVHS Review*, Fall 1993.

beyond the technical which are equally important for successful ITS deployment.

The notion of organizational readiness recognizes that ITS and related technologies enable, and perhaps force, a new, broader "enterprise-wide" view of the agency. Specifically, local transportation agencies concerned with ITS deployment need to recognize that ITS is inherently *intermodal* and *inter-organizational*.

The flow of information enabled by ITS technologies can create new relationships among local transportation organizations, among federal, state and local transportation agencies, and between public- and private-sector organizations. Indeed, our local organizations must respond to the challenge of a new level of information availability and quality. The question is no longer gathering and processing information, but knowing what to do with it once one has it.

ITS provides a local transportation organization with an opportunity to think more effectively about their "customers" in providing a wide range of transportation and information services, perhaps financing these services through innovative partnering.

CHANGE IN VIEWPOINT

ITS requires a change in the viewpoint of the local transportation organization away from building conventional infrastructure and toward *operating* the infrastructure and *connecting* to users -- i.e., customers -- through ITS technologies. The utilization of ITS for *control* of transportation flows -- such as in response to environmental emergencies -- as well as simply providing improved mobility, creates a new political challenge for these agencies. For the same reason, the deployment of "congestion pricing" will likewise require a change in outlook.

Deployment of ITS is a multi-dimensional undertaking with capability in technology, systems and institutions being needed. The talents of the "new transportation professional"[4] recruited to local agencies can help achieve this.

However, we must go further than simply hiring new people with new skills. Fundamental changes in outlook and perspective, as noted above, in our local transportation agencies are needed as well. To be successful in achieving these changes, leadership and commitment at the senior level is essential. One viable approach: agencies can recreate themselves as

---

[4] Sussman, Joseph M., "Educating the 'New Transportation Professional'", *ITS Quarterly*, ITS America, Washington, DC, Summer 1995. N.B. This article appears in Section III of this volume.

"learning organizations," as advanced by Peter Senge in his book *The Fifth Discipline*. This continuously improving organization, responding to new external factors, demands and opportunities, can develop the capabilities -- technical and beyond -- needed for implementing a successful ITS program.

All of this is not to underestimate the need for expanded technological expertise for ITS deployment; indeed, it is essential. Our message here is simply that our local transportation enterprises must substantially reshape themselves to achieve organizational readiness as information and communications-intensive, intermodal, inter-organizational, customer-oriented agencies.

IV. 2. CO-OPETITION: A FRAMEWORK FOR ANALYZING ITS RELATIONSHIPS[1]

One does not have to be around the Intelligent Transportation Systems (ITS) program for very long to realize that there are many different kinds of organizations involved. There are public-sector organizations at the federal, state and local levels, and private-sector organizations, such as automobile companies, communications companies, digital map producers, etc., who provide ITS hardware and software. The research and education community is involved, as are associations representing professional groups. Finally, we have a vast array of users (i.e., customers), including the driving public, the trucking industry, and public transportation users, among others.

Many of us have struggled from the time of the ITS strategic plan development in 1991 and 1992 to the current day with trying to understand these relationships and to have a framework for dealing with them. A recent book called *Co-Opetition*[2] provides a promising start.

The authors introduce the concept of "the value net."

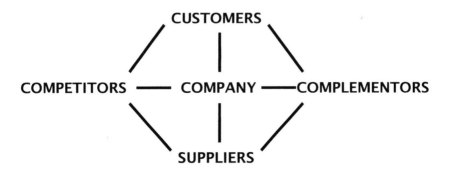

The value net suggests that every organization interacts with suppliers, competitors, customers and complementors. Suppliers, competitors and customers are familiar concepts. Complementors help the organization by making its product more attractive (e.g., hot dogs and mustard) or by expanding the market.

---

[1] Reprinted with permission of ITS America. Sussman, Joseph M., "Co-Opetition: A Framework for Analyzing ITS Relationships", "Thoughts on ITS" Column, *ITS Quarterly*, ITS America, Washington, DC, Fall/Winter 1996.

[2] Brandenburger, Adam N. and Barry L. Nalebuff, *Co-Opetition*, Currency Doubleday, New York, 1996.

The premise of the book is that each organization can use the value net as a way of structuring relationships and, in particular, can build complementary relationships with suppliers, customers and even competitors. Organizations may complement one another in "growing the pie" (in our case, the ITS pie), and compete when the time comes to split up the (bigger, we hope) pie.

The authors use transportation as an illustration of their framework, going back to the early 20th century to discuss the Lincoln Highway Association, made up of competitors, complementors and suppliers such as General Motors, Hudson, Packard, and Willys-Overland, together with Goodyear tires and Prest-O-Light headlights, working together to expedite development of America's highways, given they now had this new technology -- automobiles -- to sell.

Using the co-opetition framework in modern-day transportation can be fruitful as well, and it rings true in the ITS environment. The authors speak of "the player you can't avoid" -- the *public sector* -- noting the government can play a role as customer, supplier, competitor *and* complementor, depending on the particular context. Certainly this is true in ITS as one considers the complementary interactions of ATMS and ATIS technologies, deployed typically by the public and private sectors, respectively, and the public sector as a supplier of concepts, such as the national system architecture.

"Bringing in customers" is a key theme. Educating the market and, perhaps, subsidizing some customers is discussed as a strategy within the value net. We are doing this in ITS. For example, various ATIS systems provide subsidized, or free, traveler information to the marketplace to educate the users about the value of their services, hoping to transform them into full paying customers in the future. Indeed, the mechanism by which the private sector is able to provide that free service is sometimes through subsidies from the government, as the latter looks at opportunities for "bringing in customers" for the broader ITS program.

Co-opetition speaks of "bringing in suppliers" and "paying them to play." Establishing a network of sophisticated suppliers so that no one supplier dominates is a key to the long-term development of a technologically-based system. In ITS we did this through funding of numerous organizations, many new to the ITS area, to develop a national system architecture. The public sector saw the architecture as key to a nationally-scaled, non-balkanized ITS system, but also saw the benefit of bringing in a number of new players to "supply" the ITS industry.

Co-opetition emphasizes that even direct competitors can be complementors in some instances. In the exhibit area at the recent third World Congress in Orlando, Florida, many were struck by the large number of new players -- many with new ITS ideas -- in the ITS marketplace. I heard several Butch Cassidy-like "Who *are* these guys?" comments. These "guys" bring stronger competition for the ITS dollar, but also bring new concepts and people that can expand the intellectual capital in the ITS program. While cutthroat competition (such as we have observed in the electronic toll collection segment of ITS) serves no one's interest in the long run, the introduction of new technological players can serve all interests at this stage of ITS development and deployment.

As we move toward re-authorization of ISTEA, I believe co-opetition and the value net is a productive way of thinking about the future of ITS. We need to make sense of the complex relationships that we deal with in ITS. The environment -- that of a growth market in transportation technologies, many of which are still in the development stage -- makes our interrelationships complex.

Further, the systemic nature of the technology -- thinking of the vehicle and infrastructure as coupled, unlike in earlier systems like the Interstate, where the vehicle and highway were essentially uncoupled -- adds a critical dimension. While much ITS research and development has taken place, we are still in a nascent stage of ITS deployment. I argue that we are at the stage of "growing the pie" collectively, something of interest to the ITS community at large. Using the co-opetition framework for thinking how this growth can most effectively take place is in our collective interest. While competition certainly exists, and in many aspects of ITS is already strong, having a larger pie for us all is in our interest, and given the importance of transportation in economic development and to quality of life, in the interest of the country and the world.

IV. 3. TEACHING ABOUT ITS -- A MOVING TARGET[1]

As I write this column, I am in the midst of teaching a graduate course at MIT called "An Introduction to Intelligent Transportation Systems (ITS)". I have been teaching this course since 1993 and liken it to teaching physics in the era of Sir Isaac Newton preparing his "Principia". The field changes in real-time over a semester. I find it very interesting to look back at the subject as I first taught it in 1993 and compare it to what I am teaching now. Certainly we know a lot more answers but, at the same time, we have uncovered a lot more questions.

Back in 1993, the subject was named "An Introduction to Intelligent Vehicle Highway Systems"; the change in the name mirrors the change of the concept itself, reflecting in turn a broader perspective on what can be accomplished with these technologies beyond automobiles and highways.

Beyond the name changes, there is certainly a lot of new terminology. We now talk of CVISN (Commercial Vehicle Information Systems and Network program), ITI (the Intelligent Transportation Infrastructure), Operation Time-Saver, MDI (the Modal Deployment Initiatives) and the National ITS Goal as major deployment initiatives in the ITS sphere. Indeed, deployment is the major focus of the program, while R & D was the focus in 1993.

We have seen the establishment of 17 ITS America chapters, embracing 24 states around the U.S., reflecting the recognition that deployment will occur at the local, not federal, level.

MEASURING ITS BENEFITS

In 1993 there was no Joint Program Office (JPO) in the U.S. Department of Transportation. The establishment of the JPO certainly signals a multimodal/intermodal frame of reference for ITS, a direction urged by many. In 1993, I talked in class about how we were *going to* measure ITS benefits. Now, we finally have reliable ITS benefits information and "just-

---

[1] Reprinted with permission of ITS America. Sussman, Joseph M., "Teaching about ITS -- A Moving Target", "Thoughts on ITS" Column, *ITS Quarterly*, ITS America, Washington, DC, Spring 1997.

in-time" because in this year's class we will also talk about the ITS component of the re-authorization of ISTEA. Demonstrating ITS benefits is vital in this context.

In 1993 I talked vaguely about the public/private partnerships we needed for ITS. Now in 1997, we have a number of public/private models in advanced traveler information systems, privately-operated highway facilities and consortia of public- and private-sector organizations for research and development in automated highway systems (AHS) and other areas. In 1993, the 1997 demonstration of AHS seemed far away. This summer, we will have the opportunity in San Diego to see AHS operate on a prototype basis.

ISSUE OF BALKANIZATION

In 1993 we were just embarking upon the development of a national system architecture for ITS. Mea Culpa: I told my class in 1993 that developing this system architecture was both necessary and sufficient for developing interoperable ITS deployments around the country. This year I will tell them that, with the national system architecture having been complete for almost a year, the system architecture is certainly very useful but not really a necessary condition for interoperability, and certainly not sufficient -- absent standards -- for that purpose. Although we see a new focus on "regional architectures" derived from the national system architecture, the issue of ITS balkanization probably looms as large now as it ever has.

In 1993 the Atlanta Olympics were three years off and using ITS there was -- at best -- a gleam in someone's eye. In 1997 we will spend time talking about what we learned from the Atlanta showcase.

This year's class will include new areas of substantive research like *user adoption* of ATIS and development of major hardware/software *simulation laboratories*, enabling careful analyses and design of ITS systems. We will discuss topics like *ITS and sustainable communities*, based on a developing literature. None of these were on the agenda -- or at least in my curriculum -- in 1993. My reading list this year contains references from the three World Congresses held to-date; in the spring of 1993 we were a year and a half away from the first one in Paris.

DIVERSE STUDENT BODY

Perhaps most interesting to me and most encouraging, are the students taking the class. In 1993, the class was taken only by students in MIT's Transportation program. In subsequent years, we have had students from Urban Studies and Planning, Architecture, Mechanical Engineering, Construction Management, the Technology and Policy Program, and from the Kennedy School of Public Policy at Harvard, as well as transportation students.

Students are seeing the excitement in transportation through ITS. This development, observed not only at MIT, but at universities around the country, augurs well for the future; ITS is proving to be a pipeline from various disciplines into the transportation field. This is a basic step in the development of the "New Transportation Professional" discussed in earlier articles.

What will we be teaching in this subject in 2001, four years hence? One cannot be sure, but I can say with confidence that there will be new ideas, concepts and developments that none of us can anticipate today. We will certainly teach the fundamentals of ITS, but the future thrust of the field and its impacts on our transportation system as well as the broader socio-political-economic system and transportation are impossible to foresee. All we can say for sure is that it has been and will be a heck of a ride.

IV. 4. ITS AND SAFETY:
A WORST CASE SCENARIO[1]

From almost the earliest days of ITS in the U.S., enhanced safety has been the foremost goal of the program. Congestion relief, the leader out of the starting gate in the mid-1980s, soon lost its pre-eminent position to safety, which was viewed as a more pragmatic as well as a more ethical goal for a nationally-scaled program.

Indeed, the early evidence suggests that enhanced safety is an achievable goal. Using ITS technologies can lead to a transportation system that is substantially more safe than we have today. In a recent paper, Scott and Farber discuss ITS-related approaches to improving safety through preventive action, collision avoidance, ergonomic safety, human factors and driver support.[2] The Automated Highway System (AHS) concept is designed to enhance highway safety by taking the human "out-of-the-loop".

INTENSE SCRUTINY FOR ITS

That is the up side. But there is a down side as well. ITS technologies, viewed by the public as advanced in nature, will be subjected to intense scrutiny, particularly where safety is involved. Research in various areas, such as nuclear safety, suggests that the negative impact on a program when an accident occurs grows substantially when the system is comprised of new and advanced technologies. Any accident, particularly one with multiple fatalities traced directly or indirectly to "new" ITS technologies, will have major negative public relations impacts on the program.

Further, there is greater potential for critical, if not "fatal" damage to a program when a major accident occurs, if there is some reason for skepticism on the part of the public. One example of this is ValuJet, a low-cost airline, which had a crash in the Florida Everglades in 1996 with more than 100 fatalities. The public's predilection to believe that a cut-rate airline

---

[1] Reprinted with permission of ITS America. Sussman, Joseph M., "ITS and Safety: A Worst Case Scenario", "Thoughts on ITS" Column, *ITS Quarterly*, ITS America, Washington, DC, Summer 1997.

[2] Scott, Susan, and Farber, Eugene, "Safety in Our Hands: An Overview of Intelligent Transportation Safety and Human Factors Technologies", *Traffic Technology International*, February/March 1997.

probably takes safety short-cuts to lower costs, put together with the actual event, dealt a substantial blow to the business prospects of ValuJet and, indeed, other such airlines.

Another example: The 1993 Amtrak accident in Alabama, when a bridge knocked out of alignment by a barge in non-navigable waters led to a derailment and 47 deaths, supported the public's view of the rail industry as an old-fashioned enterprise and was a public relations debacle for Amtrak.[3]

When accidents of this magnitude occur, the effects are "non-linear" in the sense that one big accident with, say, ten fatalities has a much larger organizational impact than ten "small" accidents with one fatality each. Media attention to those accidents is intense; it reflects the ongoing reappraisal by society of whether our collective appraisal of societal risk for a particular system -- low-cost airlines, Amtrak, etc. -- is correct. It can lead to counterproductive regulatory action and can affect the viability of an organization or program.

PROFOUND IMPACT OF AN ACCIDENT

Could such a thing happen in the ITS arena? We all hope not, and the ITS community will work very hard to lower the probabilities of serious ITS-related accidents as much as possible. Nonetheless, if we are to be realistic and professional in our thinking, we must recognize that such events have a non-zero probability of occurrence, and the impact of such an event on the ITS program, particularly in its early years, could be profound.

Let me paint a worst-case scenario. Recall the 1994 Fox River Grove, Illinois, train collision with a school bus at a grade crossing. The profoundly unfortunate deaths of seven high school students in that accident reverberated around the nation for weeks. The number of fatalities, combined with who was killed -- "Our teenagers have enough ways of getting in trouble on their own without us killing them in school buses" -- had major negative impacts on the credibility of the organizations involved. This kind of accident at a grade crossing equipped with ITS technologies, say in 1999, could shake the very foundations of a nascent ITS program.

We will strive to avoid an ITS-related accident, but as professionals we must recognize that it can happen. We have a responsibility to think ahead about how to manage the program if such an event occurs. We need a plan for a pro-active approach in the event of ITS-related catastrophes; we need to assure the viability of an ITS program that can enhance long-run safety

[3] Roth, Daniel, and Sussman, Joseph, "A Risk Assessment Perspective on the Amtrak Train Crash of September 22, 1993", Report 15, Cooperative JR East/MIT Research Program on Risk Assessment, August 1994.

and other quality-of-life variables in a major way, even if a rare safety-related ITS incident does occur. We need to recognize the risks and plan for such (highly unlikely) events now. Better we have such plans in place and never use them, than fail to develop them in the misguided hope that "it can't happen to us" -- and it does.

IV. 5. AHS, ITS AND AWARENESS[1]

As I write this, I have just returned from San Diego, where I attended the Automated Highway System's Demo '97, and the ITS America Board of Directors Workshop and Meeting.

Demo '97 is, virtually by acclamation, a major success. Many (including me) were very skeptical when ISTEA, written and passed in 1991, included a provision for the demonstration of AHS in 1997.   "Could anything meaningful be accomplished in that short a time frame?" many of us asked.

Indeed, a great deal *did* get accomplished. A number of experimental concepts were unveiled, including the University of California PATH Platoon Vehicle System, the Houston Metro-Carnegie Mellon University automated free-agent bus, the Toyota Evolutionary Vehicle, and others. The range of technologies and approaches that operated without a hitch during Demo '97 was impressive. Hats off to the National Automated Highway System Consortium, under the directorship of James Billings of General Motors, for making this happen.

But, in addition to the technological tour-de-force exhibited during Demo '97, another important concept -- the evolutionary approach to AHS -- seemed omnipresent. It seemed to me that those advocating the incremental (although important) advances in areas such as advanced traffic management systems and advanced traveler information systems, and those arguing for the revolutionary breakthrough technologies of AHS, capable of more substantial improvements but in a much longer time frame, came together in San Diego.

AHS AS LONG-TERM INTEGRATED APPROACH

The debate, rather than being a divisive battle for resources between the incrementalists and the revolutionaries, was recast as the community came together to think of AHS as a part of a long-term integrated approach to ITS. The linkages of AHS technologies to both the short-term and long-term goals of ITS were manifest in the activities at Demo '97.

---

[1] Reprinted with permission of ITS America.   Sussman, Joseph M., "AHS, ITS and Awareness", "Thoughts on ITS" Column, *ITS Quarterly*, ITS America, Washington, DC, Fall 1997.

This joining of the ITS and AHS communities is timely because of the clear need to build *awareness* of ITS. Recent articles by Bob Carr[2] and Harry Voccola[3], current Chairman of the ITS America Board of Directors, highlight the importance of awareness -- in several constituencies. We need to make appointed and elected officials (and associated staff) at all levels of government aware of the benefits of deploying ITS infrastructure. Further, the general public and business interests, which represent both the mechanism to build political support for ITS investments as well as a huge market for in-vehicle equipment, must be made aware of ITS and the benefits that will accrue to individuals and businesses. Awareness must be at the heart of any strategy to build the ITS program.

POTENTIAL FOR DEVELOPING AWARENESS

Demo '97 showed the potential for developing awareness, with a considerable amount of coverage in the national and international print and electronic media. The lesson is that when we have something newsworthy to say, the media will pay attention. And the notion that both short- and long-term benefits -- to society and individuals -- are achievable through the deployment of advanced transportation technologies is an important message to convey.

We need to make people knowledgeable about ITS at several levels. For the professional community, DOT's program of Professional Capacity-Building (PCB) (see Christine Johnson's article in this issue of the *ITS Quarterly*[4]) is required for the development of an organizational infra-structure that can deploy the transportation/information infrastructure em-bodied in ITS. Graduate education, producing the "new generation of trans-portation professionals",[5, 6] represents the longer-term approach to building professional infrastructure. But knowledge of ITS benefits -- awareness -- must extend to our appointed and elected public officials and to the general public if long-term political and market support is to be developed.

---

[2] Carr, Bob, "ITS: Success Requires Public Awareness", *ITS Quarterly*, ITS America, Washington, DC, 1996 Fall/Winter Issue.

[3] Voccola, Harry, "Why You Should Join the National ITS Awareness Campaign", *ITS Quarterly*, ITS America, Washington, DC, 1997 Summer Issue.

[4] Johnson, Christine, "ITS: The Five-Year 'Experiment': A Progress Report", *ITS Quarterly*, ITS America, Washington, DC, Fall 1997.

[5] Sussman, Joseph, "Educating the 'New Transportation Professional'", *ITS Quarterly*, ITS America, Washington, DC, Summer 1995. N.B. This article appears in Section III of this volume.

[6] Benson, Brien, "ITS Education in American Universities", from "A Symposium on Education", *ITS Quarterly*, ITS America, Washington, DC, Fall/Winter 1996.

I chair an ITS America Board of Directors panel charged with overseeing the ITS America awareness campaign, with the help of the Communication and Outreach Task Force chaired by Hap Carr. The panel was gratified to see the Board of Directors rank awareness as Number 2 on a long list of priority areas for ITS America (second only to standards development) at its San Diego workshop.

This emphasizes that pushing the awareness agenda is a critical priority for all of us in the ITS community. The AHS Demo '97 has given us a good model and an important leg up on doing exactly that.

IV. 6. THE ITS ROLE AT THE MILLENNIUM[1]

In September 1997, the prestigious "Annals of the American Academy of Political and Social Science" published a special issue entitled "Transport at the Millennium". Comprised of 18 essays, it provides a broad perspective on the transportation field: where it has been, where it is now, and where it is likely to go in the future. Freight and passenger transportation in all modes are discussed, along with experiences in developing countries and Europe.

As I read this set of essays, I was struck by what seemed to me to be a real paradox. While some of the fundamental themes of this special publication focus on areas in which Intelligent Transportation Systems (ITS) can be a major factor or enabler, ITS was hardly mentioned as an important factor in current-day and future transportation systems.

In the preface to the volume, Stanley G. Long, the Special Editor for this issue, notes that two themes recur again and again in the essays. These two themes are:

1. A fundamental change in the relationship between the public and private sectors in the transportation industry, embodied in deregulation -- let the market work -- and privatization -- the notion that transportation infrastructure as a publicly-provided good may not be in accord with today's market-driven economies.
2. Market failure in transportation because of unpriced externalities caused by the operation of transportation systems. Included here would be congestion, environmental impacts, etc.

Dr. Long is quite correct. As one reads the essays, these themes do indeed recur again and again. The notion of letting the transportation system operate within a market framework, including pricing of externalities, is viewed by many of the authors as of fundamental importance in our increasingly competitive and environmentally-conscious world in passenger and freight services considering all modes of transportation and intermodalism.

---

[1] Reprinted with permission of ITS America. Sussman, Joseph M., "The ITS Role at the Millennium", "Thoughts on ITS" Column, *ITS Quarterly*, ITS America, Washington, DC, Winter 1997/Spring 1998.

PUZZLING DISCONNECT

Puzzling to me as an ITS educator and researcher is the disconnect between the set of articles advocating these views, and the potential of ITS for serving these ends.   ITS technology can be an important enabler of private operation of transportation systems components traditionally in the public sector, most notably in the highway field. Technologies can be used to permit collection of revenues at the point of use by private transportation providers and, importantly, without interfering with traffic flow. Private-sector initiatives in providing real-time transportation information represents a new and potentially huge market in the U.S. and abroad.

ITS technologies also provide the basis for pricing of transportation services so as to make the cost of externalities explicit to the user. Road pricing or congestion pricing is the classic example of having road users pay for the congestion impacts they cause, potentially limiting temporal peaks in traffic. Extending these strategies as a mechanism for improving air quality is potentially a viable idea as well.

So, I found "Transport at the Millennium" both disappointing and exciting. It was disappointing, of course, in the sense that the potential of ITS is not explicitly recognized. It was exciting in the sense that the key issues identified by a number of transportation scholars -- the change of the relationship of public and private sector in market-driven transportation, and the importance of recognizing and pricing transportation externalities -- are at the heart and soul of what ITS can provide.

ARTICULATING LONGER-TERM VISION

ITS can be the mechanism through which market-driven, intermodal, globally-scaled transportation enterprises can operate and flourish.   But despite the outpouring of literature on ITS -- and the magazines and journals are large in number -- this forward-looking message is not getting out. ITS as simply a reducer of congestion in urban areas and as the purveyor of gee-whiz electronics for one's vehicle is the image we present.   People in leadership positions in the public and private sectors and in the research/ academic community need to be more effective in articulating this longer-term vision of ITS and what it can provide to the world of transportation and the changing social/political/economic environment in which we operate.

The lead article in "Annals" is entitled "Transportation Today: The U.S. Experience in a World Context", by John R. Meyer, a distinguished transportation economist from Harvard. Professor Meyer states, "In fact, like so many other sectors of the economy, transportation is increasingly becoming a thinking person's business, with shorter technological and product life cycles requiring increased ability to adapt quickly and flexibly." This is the working philosophy of ITS, and we must represent it as such to decisionmakers and the community-at-large.

IV. 7. ITS AND THE FEDERAL TRANSPORTATION SCIENCE AND TECHNOLOGY STRATEGY[1]

Earlier this summer I chaired a Transportation Research Board (TRB) panel to review the Federal Transportation Science and Technology strategy, which has been in development over the past year. The strategy currently consists of eleven "partnerships" designed to bring together players from the federal government, state and local governments, the private sector and the academic research community.

There is a lot of good news in these partnerships for the ITS community. No less then five of the eleven are ITS-related.

A partnership on "Accessibility for Aging and Transportation-Disadvantaged Populations" deals with providing mobility through better transportation management and advanced technologies, including automatic vehicle locations, computer-aided dispatch and electronic toll collection, all parts of the ITS program.

A partnership on "Enhanced Goods and Freight Movement at Domestic and International Gateways" focuses on information exchange and technologies designed to enable innovative logistics. It has aspects that are at the heart of ITS.

Further, a partnership on "Transportation and Sustainable Communities" deals with the balance among economic growth, environmental quality and sustainability, which is a core principle of ITS. These goals are to be attained through better forecasting, planning and assessment tools enabled by ITS-gathered data.

HEART OF the FEDERAL INITIATIVE

And then, of course, we have the twin towers of the ITS Federal program -- the "Intelligent Vehicle Initiative (IVI)" and the "National Intelligent Transportation Infrastructure (NITI)". IVI and NITI, based on modern

---

[1] Reprinted with permission of ITS America. Sussman, Joseph M., "ITS and the Federal Transportation Science and Technology Strategy", "Thoughts on ITS" Column, *ITS Quarterly*, ITS America, Washington, DC, Summer 1998.

technological developments in information technology (IT) and communications, comprise the heart of the federal ITS initiative; and, in accord with the Science and Technology approach, both are certainly and necessarily partnerships.   IVI deals with the application of advanced technologies to in-vehicle driver assistance systems and clearly requires a partnership between government and private-sector automobile manufacturers.   NITI has been characterized as a "communications and information backbone", enabling ITS products and services to work together. The NITI partnership, involving an infrastructure composed of an integrated set of advanced technologies, reflects the clear historical federal role in infrastructure provision, again in partnership with state and local governments.

So the good news is that ITS is certainly in the game in terms of the federal transportation science and technology strategy, with almost half the partnerships involving ITS directly and with the two key initiatives of ITS -- IVI and NITI -- being among the areas highlighted by the federal science and technology program.

While this picture is bright, there is an important concern reflected in the characterization of IVI and NITI as two distinct partnerships. Certainly there are pragmatic reasons for this, given some of the institutional and market issues that exist, along with the fundamental differences between IVI and NITI in the relevant technologies and in the ways costs and benefits accrue.   But it is important that we not lose sight of the "fundamental insight" of ITS; this is the *linkage* between the vehicle and infrastructure components of the transportation system.

The electronic and information linkage between vehicle and infrastructure is the major point of departure from the traditional "interstate" transportation perspective; many potential advantages of ITS flow directly from that linkage. The "separation" of IVI from NITI suggests a system that may be less than fully integrated. In the development of these two science and technology streams, we should ensure that no barriers to long-term integration of vehicle and infrastructure are erected. Otherwise, some of the important safety, mobility and productivity benefits of ITS will disappear and the cost and level-of-service advantages of an integrated system will be dissipated.

CHANGES AT ITS AMERICA

Since my last column, major changes have taken place at ITS America. Jim Costantino, the first president and chief executive officer of ITS America, has been succeeded by John Collins. I would be remiss not to acknowledge the extraordinary contribution that Jim made in building ITS America (then IVHS America) from nothing (no people, and a financial deficit) into the pillar in the ITS program that ITS America is today. I had the good fortune to be at ITS America for nine months in part of the formative period during 1991 and 1992 and, I can assure you, Jim's accomplishment in building the organization from scratch was truly remarkable.

Further, I wish John Collins every success as the new CEO. I served on the search committee that led to John's appointment at ITS America and know that he is an insightful and energetic professional who brings a tremendous arsenal of capabilities to the organization. I have every confidence he will make a great leader of ITS America and the ITS community at large, as we face the challenges the future is sure to bring.

We all are fortunate to have had the services of Jim Costantino and, now, the services of John Collins as our CEOs at ITS America.

IV. 8. ITS AND "RESCUING PROMETHEUS"[1]

A new book about engineering systems called *Rescuing Prometheus*[2] has important lessons for the ITS community.

Thomas P. Hughes, the author, is Emeritus Mellon Professor of the History and Sociology of Science at the University of Pennsylvania, also a distinguished Visiting Professor at MIT, and probably the foremost historian of technology currently active. In his book, he analyzes four major technology-based engineering systems projects: the Sage Air Defense Project, the Atlas Intercontinental Ballistic Missile Project, the Arpanet (the precursor of today's Internet), and the Boston Central Artery/Tunnel Project.

His careful study of these large design and development projects, each with a heavy technological component, leads him to conclude that their common feature is the *management of complexity*. Indeed, he suggests that despite the cutting-edge technological components of each of these four projects, their most notable aspect was the ability of *system builders* to manage the extraordinarily multi-dimensional complexity that characterized each of them. He concludes that these projects have "created new styles of management, new forms of organization, and a new vision of technology".

Hughes distinguishes between the earlier "modern" Sage and Atlas projects and the subsequent "post-modern" Arpanet and Central Artery/Tunnel projects. In this context, by "modern" he means the strongly hierarchical "command and control" structure that characterized the management of Sage and Atlas, as opposed to the much flatter, distributed "post-modern" management structures used in the Arpanet and the Central Artery/ Tunnel projects.

The ITS community has much to learn from Hughes' analyses of these four engineering systems and, in particular, the latter two. Certainly, *managing complexity* is at the heart of the ITS program as we seek to deploy advanced technologies around the country and around the world to fundamentally change the quality and quantity of transportation services we provide.

---

[1] Reprinted with permission of ITS America. Sussman, Joseph M., "ITS and 'Rescuing Prometheus'", "Thoughts on ITS" Column, *ITS Quarterly*, ITS America, Washington, DC, Fall 1998/Winter 1999.

[2] Hughes, Thomas P., *Rescuing Prometheus*, Pantheon Books, New York, 1998. (*Prometheus Bound* by Aeschylus, a Greek dramatist, characterizes Prometheus as the bringer of fire, civilization, and the arts and sciences to the human race.)

ITS could be characterized as post-modern, in that its management structure is quite flat and distributed. On the public-sector side, it has become increasingly clear that deployment of ITS is a locally-oriented process. An example of this is the development of *regional architectures* from the ITS national architecture as a necessary condition to advance the program at a subnational (i.e., regional) scale. On the private-sector side, successful new technologies are most likely to be effectively developed by a market-oriented process rather than a top-down directed structure, as witness the various ITS trade shows in the U.S. and abroad.

In work at MIT in engineering systems, we have coined the term "*CLIOS*", an acronym for *Complex, Large-Scale, Integrated, Open Systems*. We view CLIOS as an important area of study; indeed, the work of Tom Hughes heralds the study of CLIOS as a serious intellectual enterprise. We argue that developing *fundamental design principles* for CLIOS represents the major engineering challenge for the early 21st century.

So, why is all this of relevance to the ITS community? Because, ITS *is* a CLIOS and we can benefit from the study of this phenomenon. Consider the definition of each word in CLIOS and its connection to ITS.

A system is *complex* when it is composed of a group of related units (subsystems), for which the degree and nature of the relationships is imperfectly known. Its overall behavior is difficult to predict, even when subsystem behavior is readily predictable. Further, the time-scales of various subsystems may be very different (as we can see in transportation -- land-use changes, for example, vs. operating decisions).

CLIOS have impacts that are *large* in magnitude, and often *long-lived* and of *large-scale* geographical extent.

Subsystems within CLIOS are *integrated*, closely coupled through feedback loops.

By "*open*" we mean that CLIOS explicitly include social, political and economic aspects.

Often CLIOS are counterintuitive in their behavior. At the least, developing models that will predict their performance can be very difficult to do. Often the performance measures for CLIOS are difficult to define and, perhaps even difficult to agree about, depending upon your viewpoint. In CLIOS there is often human agency involved.

ITS is clearly a CLIOS; indeed, it is a classic contemporary example. Certainly, it is complex, with myriad integrated subsystems. ITS is often regional in scale, and large in impact, and we have learned that explicit consideration of social, political and economic factors is fundamental to success in ITS deployment. Further, modeling and defining ITS performance is a continuing challenge.

It is useful to think of ITS as a CLIOS because then the field can benefit from the knowledge base developed in the design of other CLIOS, for example, the ones discussed by Professor Hughes. The design principles that we can abstract from careful study of and practical experience with CLIOS can be useful in the ITS context. Further, ITS can contribute to a broader understanding of CLIOS by documenting the hard-won lessons we have learned and will continue to learn in the future as ITS is deployed and operated. Through this process ITS itself gains as it matures.

IV. 9.   ITS AND CONGESTION[1]

As I write this column, we have just concluded a major conference at MIT sponsored by the Institute and Ford Motor Company. This conference, "Traffic Congestion: A Global Perspective", focused on congestion and strategic approaches to ameliorating it. The conference was international in content, with speakers discussing the developed and developing world in the Americas, Europe and Asia.

These conferences are valuable to the ITS program because they bring together practitioners and scholars concerned with broad transportation policy and operating issues, who may have a relatively modest, if any, connection to ITS. It gives us new perspectives as people deal with issues that we are concerned with from a point of view we rarely hear at ITS meetings.

So, for example, we heard at this conference that the improvement of congestion, a critical foundation of the ITS movement, is not really such a big deal. Some researchers in Europe have estimated that the impact of congestion in developed countries in Europe is probably closer to .2% of GNP rather than the 2% of GNP often cited. Some have suggested that if one does a rational analysis of congestion, its criticality compared with other societal issues goes down dramatically.

Others talked about congestion pricing. Several speakers, myself included, noted that the theoretical roots of congestion pricing originally go back to the 1920s. In the 1950s Nobel laureate William Vickery suggested that pricing of road services could move us beyond the Soviet-style queuing method for allocating roadway capacity. Now ITS is here, finally, with the technology to deploy exactly these kinds of measures.

But several at the conference suggested that the benefits of congestion pricing were ephemeral, probably inequitable from an income-distribution viewpoint, and perhaps not even cost-effective given the investments in hardware and software that one would have to make to deploy this concept. In effect, these speakers were saying, "Be careful what you wish for", in this case, technology to deploy congestion pricing techniques, because these techniques might well prove to be counterproductive and non-cost-effective, to say nothing of being politically divisive. Their not unreasonable evidence is the lack of successful congestion pricing deployments.

[1]   Reprinted with permission of ITS America. Sussman, Joseph M., "ITS and Congestion", "Thoughts on ITS" Column, *ITS Quarterly*, ITS America, Washington, DC, Spring/Summer 1999.

But there were positive views as well. The message was reinforced that a long-term strategic view of transportation, enhanced by advanced technology, could provide a more mobile society. Joseph Yee, from the Land Transport Agency of Singapore, spoke eloquently about Singapore's long-term approach to transportation, with a multi-dimensional program involving: 1) integrated land-use and transport planning; 2) the provision of roadway infrastructure; 3) the harnessing of technology and, in particular, transport telematics; 4) the management of demand through area licensing schemes and electronic road pricing; and 5) improved public transportation. The message from Mr. Yee: a multi-dimensional transport strategy can be quite effective and congestion pricing can work as part of a menu of transportation strategies.

We heard again that there is "no silver bullet" in attacking mobility and accessibility questions in modern society -- there is no single answer; rather, an organized strategic multi-dimensional approach is required. Ottawa, Canada, and Curitiba, Brazil -- exemplars of strategic transportation planning -- reflected this same idea: we must think in the long term and on many dimensions if our urban regions are to remain vital and productive economic centers with good mobility and accessibility.

In the end, the conference tells us that transport and related issues in environment and land-use planning remain at the forefront of interest in both the public and private sectors, as Mort Downey, Deputy Secretary of U.S. DOT, emphasized in his keynote remarks. For those of us in the ITS movement, this highlighting of transportation further emphasizes that the world is watching as we strive to deploy systems. ITS needs to make its mark as a contributor to improving the transport system in a manner that is consistent with approaches to other major societal questions, such as economic growth, environmental concerns and social equity.

But, what of the doubts about the criticality of congestion on society's agenda and the usefulness of pricing for highway infrastructure enabled by ITS? There will always be such doubts. We should let the political process and the market dictate how important congestion improvement is to the general public and individuals and how acceptable and useful roadway pricing is.

We ought not be distracted by doubts about the efficacy of ITS; rather, we should strive, through our continued emphasis on carefully designed deployments, to demonstrate and document that ITS can be relevant in solving the transportation issues of the 21st century.

IV. 10. REGIONAL ITS ARCHITECTURE CONSISTENCY: WHAT SHOULD IT MEAN?[1]

Back in 1994, the United States Department of Transportation (U.S. DOT), urged on by ITS America, began the development of a *National ITS Architecture*. The motivation for creating a National ITS Architecture was that the very complexity of ITS required an architecture to describe how its various elements or subsystems would communicate and what functions would be performed by each. Given especially the fact that ITS would operate nationally and would ultimately be developed and designed by both public- and private-sector organizations -- the former concerned primarily with infrastructure and the latter concerned primarily with vehicles and in-vehicle equipment -- an architecture to describe the communications links and functionalities was thought essential.

Further, the National ITS Architecture was advanced as an important step toward a *nationally-interoperable* system. It was said that the development of the national architecture would guide the development of standards, which would in turn assure national interoperability.

This concept of national interoperability has been in the thoughts of ITS advocates and designers since the earliest days of Mobility 2000 and the 1992 "IVHS" America Strategic Plan. The concern voiced then was that "balkanization" of ITS systems, developed free-form around the country, would prevent vehicles that went beyond their own regional boundaries from interacting effectively with the infrastructure in other regions. This was a particular concern to the national trucking industry, which had visions of carrying 15 transponders in their cabs to allow interaction with various infrastructures as they traversed the country. But further, this was a concern of ITS suppliers who wished to sell hardware and software into a large, integrated, nationally-scaled market.

The National ITS Architecture was developed and delivered in 1996 by various organizations under contract to U.S. DOT. The result was a substantial and content-rich set of documents which, at the cost of $20 million, certainly dwarfs the magnitude of any other studies on ITS, and

---

[1] Reprinted with permission of ITS America. Sussman, Joseph M., "Regional ITS Architecture Consistency: What Should It Mean?", "Thoughts on ITS" Column, *ITS Quarterly*, ITS America, Washington, DC, Fall 1999.

even the cost of some ITS deployments. This architecture is a major contribution to the definition of the structure of ITS and "...has served as the model for the architecture of ITS infrastructure in several European and Asian nations."[2]

The next step is the development of "regional architectures". These regional architectures, developed by various non-federal transportation organizations, were to be "consistent" with the National ITS Architecture, and this consistency would facilitate the earlier-stated goal of national interoperability.

Now, while it would appear to be virtually un-American to be against national interoperability, it turns out that this is a controversial topic. We all know that ITS deployment will take place on a *regional* or *metropolitan* scale. Will the organizations responsible for that "local" deployment be willing to be "consistent" to assure this goal of national interoperability, especially if it is more costly to do so, or if they see themselves as constrained in their deployment strategies?

Of course, the U.S. DOT is not without influence here. To quote an "Architectural Consistency" transparency from a current FHWA-sponsored Public Technology, Inc. (PTI), class: "The Transportation Equity Act for the 21st century (TEA-21), Section 5206 (e), requires the Secretary to ensure conformity with the National ITS Architecture and Standards and, further, ITS projects implemented with funds from the highway trust fund (including mass transit account)." Or, simply put, if you use federal funds, you play by federal rules.

But what will these rules say? The rulemaking in this area is an ongoing process and we are perhaps one year away from a definition of what "architecture consistency" truly means. So far, as best as the author can glean, the idea of a regional architecture will be judged consistent (or not) with the National ITS Architecture based on the *process* by which it was developed and not by its content. That is, if all the appropriate stakeholders in a region are involved in the development of the regional ITS architecture and all points of view are taken into account, that architecture will be viewed as consistent.

The idea of having a variety of stakeholder views on the table when ITS is discussed is a good one. That the regional process is inclusive in the region is certainly of value in producing a *regionally-integrated* ITS deployment. However, it has little to do with the concept of a nationally-interoperable ITS system, and does nothing to assure that equipped vehicles can work with various regional systems, or that national markets for ITS hardware and software will be established.

---

[2] Smallen, David, "How Transportation Systems Talk to Each Other", *Public Roads*, September/October 1999, pp. 2-6.

Certainly politics is critical here; the concerns that states, regions and metropolitan areas have about being dictated to by the federal government are well-known. The history of the republic has been shaped by the tension between strong central government and the prerogatives of the states. This idea of how "consistent" a regional ITS architecture should be with the National ITS Architecture is just another manifestation of this tension.

Considerable funds were expended on the National ITS Architecture and some very valuable content and structures were the result. Currently, many efforts are ongoing to develop regional architectures, including the extension of this idea beyond technology to include institutional design at a regional scale. It makes sense to build on these efforts by developing rules for consistent regional architectures that are based on their content, not simply the process by which they were produced. Content-based consistency rules can channel ITS toward national interoperability, which is of value to all of us. Yes, assuring that the process is inclusive of all regional actors is an important goal and, in fact, not all that easy to achieve. But as a definition of regional architecture consistency, it is not enough.

We need to take another harder look at the question of what consistency of a regional architecture with the National ITS Architecture means, and try to establish content-based rules that do not overly constrain the regions, but support ITS national operability and the benefits that result.

IV. 11. A CAUTIONARY NOTE[1]

Most observers of the ITS scene in the United States would say that times are very good. The Clinton administration budget announced in early February has a substantial proposed increase in funding for ITS. In November, at the ITS World Congress in Toronto, Canada, the U.S. contingent was very well represented at a very well-attended meeting. The ITS America Annual Meeting in Boston in May 2000, highlighted by "the biggest ITS project in the world", the Central Artery/Tunnel Program, locally known as "The Big Dig", is similarly shaping up in a very positive way, with a number of important sessions, panels, and a sold-out exhibition area. The Institute of Transportation Engineers has initiated a National Steering Committee, focusing on transportation operations issues engendered by ITS and the changes this will enable in our industry. The recent FCC ruling on spectrum allocation was very favorable to ITS. Indeed, ITS is seen as a mechanism for surface transportation to move from its traditional "one size fits all" service perspective to a more tailored customer service viewpoint.

Things are certainly flowing our way. So, conforming with the management maxim that the best of times are exactly when one should look most carefully for emerging strategic questions, let me mention a few issues which bear watching by the ITS community. Some of what I will mention may ultimately turn out to be positive for ITS; some may turn out to be problematic; indeed, some may be of modest impact. In any case, we ought to be thinking about the implications of these potential changes for our ITS programs.

PUBLIC SECTOR

We are in the midst of the primary season in U.S. presidential politics. Regardless who is elected president, there will be a change in administration and in the executive branch.

I would suggest that ITS has been uniquely fortunate in the current administration. Transportation leaders are supportive in much of what we want to do. Secretary Rodney Slater has clearly staked out research and

---

[1] Reprinted with permission of ITS America. Sussman, Joseph M., "A Cautionary Note", "Thoughts on ITS" Column, *ITS Quarterly*, ITS America, Washington, DC, Winter 2000.

technology as critical elements of his DOT strategy. Deputy Secretary Mort Downey has followed through with these programs, with a great deal of support for ITS. FHWA administrator Ken Weigel has been a consistent supporter of the new technological thrusts inherent in ITS and has worked purposefully toward the restructuring of the FHWA needed to propel that agency into the next century. Dr. Christine Johnson of the Joint Program Office has been an important leader in the ITS world. It would be hard to imagine a more supportive *leadership* cadre in future administrations.

So, a question: will the federal executive branch support for ITS slip back to business as usual with the change in leadership after the 2000 election?

We also face a possible change in the balance of power in the House and Senate as well. While surface transportation has traditionally had a bipartisan flavor, nonetheless, new chairs on key congressional committees introduce risk *and* opportunity.

INDUSTRY ISSUES

On the industrial scene, the automobile industry is going through an important consolidation on an international scale in response to what are seen to be over-capacity problems in that industry. How all this will play out is far from clear. We have already seen international mergers, DaimlerChrysler, for example, and others likely to come. How this will affect ITS and even whether it will be for better or for worse is hard to predict. It is, however, a fundamental change in an industry underlying the ITS initiative.

On the information technology front, Microsoft's MSN consumer thrust on the Internet and its focus on "end-user productivity" has the potential for a major impact on ITS. However, at the same time, the Microsoft antitrust case is moving through the judicial process. Judge Thomas Penfield Jackson, in his findings of fact, signaled a ruling that could lead to a substantial restructuring of that software giant. Indeed, rumors abound about Microsoft entering into a consent decree, which might well have that same restructuring effect. The implications for ITS? Hard to say, but a fundamental restructuring of the software industry and, more generally, the information technology industry, a foundation of ITS, may well have important implications.

All of this is not to take a pessimistic view of ITS' future. I personally am quite optimistic. It is to suggest that in our dynamically changing world, several cornerstones may well be shifting, perhaps favorably, but perhaps not. We owe it to ourselves to be thinking through the implications of these changes before they are upon us, and outline strategies for ITS that can deal with potential important changes in our political and industrial environment.

IV. 12. IT HAPPENED IN BOSTON[1]

As I write this column, ITS America is fresh from its 10th anniversary Annual Meeting here in Boston, held in the first week of May. Everyone sees a complex event like the Annual Meeting through his or her own personal lens, and I am no exception. From my perspective as Program Chairman of that meeting, I feel that we made clear our standing as *the high-quality provider of ITS information in the world.*

This year we instituted a more rigorous review policy for technical presentations and, in line with the strong emphasis on deployment in ITS, we focused our attention on what has actually been accomplished, as opposed to what is going to be done in the future. I hope you will agree that we presented a high-quality technical program that reflected this deployment focus.

Again, through my personal lens, there were two aspects of the Annual Meeting that struck me as particularly interesting.

First, and again in the spirit of our deployment emphasis, there was a strong focus on the media and media relations at the Annual Meeting. I chaired a session in which we turned the tables and questioned representatives of the media on how best to put the ITS story before them. People from various print media participated: Laura Brown from *The Boston Herald*, a daily newspaper; Chana Schoenberger from *Forbes*, a business bi-weekly magazine; William Angelo, *Engineering News-Record*, the magazine of record in the construction industry; and William Carey, *New Technology Week*, a high-tech weekly newsletter. Each gave their perspective on how ITS is seen by *their* readership and what would make a story newsworthy to them. From them we learned that the "lives/time/ money" phrase introduced by our president John Collins is quite useful to catch the public's attention, but that "intelligent transportation" is, as noted by one of our panelists, "inherently unfriendly". Focusing on *results*, not the political process, was further advice.

A closing highlight of the conference, again emphasizing media relations, was a press conference format session. Five members of the ITS America Board, including our outgoing and incoming chairs, Joseph Giglio of Hagler Bailly and Northeastern University, and Harold Worrall of the Orlando-Orange County Expressway Authority, and our incoming vice-chair,

---

[1] Reprinted with permission of ITS America. Sussman, Joseph M., "It Happened in Boston", "Thoughts on ITS" Column, *ITS Quarterly*, ITS America, Washington, DC, Spring 2000.

Lawrence Yermack of PB Farradyne, were questioned by media representatives. There were several spirited exchanges between our Board members and the media and, again, we got a sense of the kinds of issues the media would like to probe about ITS and its applications. Of particular note was the interest of the media in public transportation and what ITS could contribute there.

Another aspect of the Annual Meeting that struck me as having particular relevance was the focus on *regional transportation issues and ITS*. As background, ITS America has set up a New Regional Organizations Task Force, chaired by Professor Jonathan Gifford of George Mason University, which includes regional practitioners such as Matthew Edelman of Transcom. This task force, part of the Societal Institutional and Environment Committee, sponsored a technical session with various papers on regional issues and ITS, and while I was pleased to see that session on the program, I must confess I was not optimistic about how popular it would be. So, it was a quite pleasant surprise to see more than 70 attendees in the audience to hear a set of excellent papers focused on how ITS can advance through the regional transportation agenda and how, in turn, the focus on regional transportation can accelerate what we are trying to do in ITS deployment.

I have long argued that this is a natural partnership. Indeed, in the very first of these columns that I wrote back in 1996, I said that

> *The strategic vision for ITS is as the integrator of*
> *transportation, communications and intermodalism*
> *on a regional scale.*

I think this regional emphasis is quite healthy and important. ITS provides the mechanism for us to manage transportation at a regional scale for the first time. Distinguished thinkers, such as Michael Porter at the Harvard Business School, emphasize that regions are the basis of competitiveness within the global economy. Often it is at a regional scale that we are concerned about measuring and managing environmental impacts of transportation. And regional architectures are gaining prominence as an avenue for effective regional strategic transportation planning. So it is good to see this regional ITS focus has a strong constituency in ITS America.

It was a pleasure and an honor to serve as Program Chair for the 2000 Annual Meeting in my home city of Boston, and I wish the best to the Florida organizing committee as they point toward the next Annual Meeting in Miami in June 2001.

IV. 13. CONSIDERING ITS AS A COMPLEX ADAPTIVE SYSTEM[1]

1.   INTRODUCTION

Last spring I had the opportunity to spend several weeks at the Santa Fe Institute (SFI) in New Mexico. The province of this fascinating place -- which includes among its founders Physics Nobel Laureate Murray Gell-Mann -- is *complex adaptive systems* (*cas*), systems whose behavior is very difficult to predict because of complex interactions among the various components. The scientists (physical, social and political) at Santa Fe Institute call those components *adaptive agents*, which could include people, "smart" algorithms, and organizations.[2] These *cas* are often found in nature -- evolutionary biology is a good example -- but my idea in visiting SFI was to see if the concepts, philosophies and mathematics of *cas* could be applied to the world of transportation and, in particular, Intelligent Transportation Systems (ITS).

The best application I could develop was from the most venerable aspects of ITS, namely, Advanced Transportation Management Systems (ATMS) and Advanced Traveler Information Systems (ATIS, the focus of this issue of the *ITS Quarterly*). While we have been working on these systems for some years, one could imagine higher-performance systems, advanced by a better *integration* of ATMS and ATIS concepts. Integration to improve performance is a classic system design principle; but complexity often results from that integration.

2.   INTEGRATED ATIS/ATMS

What do we mean by "better integration"? Consider an ATMS which is trying to predict congestion as it may occur in the future, based on the current state of the network and some expectation of additional traffic

---

[1]   Reprinted with permission of ITS America. Sussman, Joseph M., "Considering ITS as a Complex Adaptive System", "Thoughts on ITS" Column, *ITS Quarterly*, ITS America, Washington, DC, Summer 2000.

[2]   Holland, John H., *Hidden Order: How Adaptation Builds Complexity*, Perseus Press, September 1996.

coming into that network over some relatively short-term time horizon -- say, ten minutes. The ATMS is charged with developing strategies that will try to prevent this predicted congestion from occurring and allow the network to operate closer to optimum performance. Those strategies would include well-known ideas, such as ramp metering, dynamic traffic signal adjustment, speed limit changes, lane use changes, and Variable Message Signs giving instructions to drivers.

In addition, the ATMS can suggest routes, through traveler information, in-vehicle to drivers currently on the system. One could imagine sometime in the future that this information could be personalized to individual drivers because *the ATMS knows where the individual drivers want to go*. For example, I drive to MIT in the morning while someone else is driving to Harvard (why, I cannot imagine). Although we are both taking Route 2 in from the western suburbs at the same time, the dynamic routing suggestions provided to me as congestion develops would be different than the information provided to my Harvard colleague, because of the difference in our destinations.

So the ATMS, knowing what strategy it plans to implement to reduce network congestion, provides routes to drivers via an ATIS using that foreknowledge, as well as its knowledge of where drivers are going.

But, of course, it is *not* that simple. We do not know whether the drivers will accept that routing advice. I may think I can outguess the system and get to MIT more quickly via a route I have chosen based on experience -- after all, I am an adaptive agent, in the parlance of the Santa Fe Institute. So the ATMS, in deciding what the optimal strategy is for the next ten minutes on the Boston metropolitan network, would have to take into account the possibility that individual drivers may choose routes other than the ones suggested by the ATIS. Indeed, the ATMS is an adaptive agent as well and can learn over time what drivers will do. And, in fact, what these drivers choose to do depends on their judgment about the quality of routing information being provided, which in turn depends on how effective the ATMS is at "guessing" what drivers will do, ... well, you get the idea.

My colleagues at the Santa Fe Institute assured me this integrated ATMS/ATIS concept fits the classic definition of a complex adaptive system, and therefore would be subject to all the issues facing such systems -- counterintuitive emergent behavior, instability and non-linearity in performance, and so forth. But also, by characterizing ITS as a complex system, we have open to us a whole set of methods developed for *cas*.

3.   COMPLEXITY AND INTEGRATION

Current ITS applications are often independent, one from the other. For example, in many cities the electronic toll collection application runs independently of the ATMS in that same metropolitan area. The way we will achieve a superior level of performance in the next ITS generation is to better integrate applications, such as the integrated ATMS/ATIS described above.

But, while integration is needed to improve system performance, system complexity often occurs as well.   So, in designing these integrated applications, knowledge and insights developed by scientists dealing with complex adaptive systems in the natural world can perhaps be applied to ITS as an "engineering system", where human cognition and technology interact.

Virtually since its inception, ITS has succeeded in the world of transportation by reaching out to new technologies and concepts,. Using the concepts and methods of *complex adaptive systems* to create integrated and more effective ITS applications is a next step in that tradition.

When you read the many excellent ATIS articles in this issue, perhaps it is instructive to think about how those systems might operate more effectively if one considered integration, complexity and adaptability in their design in the next generation of ITS.

IV. 14. MEGA-CITIES IN DEVELOPING COUNTRIES -- A MAJOR ITS MARKET FOR THE FUTURE [1]

As I pen this column, the "Subway Series" -- the World Series between the New York Yankees and the New York Mets -- is underway. "Subway" is, of course, a transportation term -- you can get from Yankee Stadium to Shea Stadium on the New York subway.

The last such series took place in 1956 between those same New York Yankees and the late, lamented Brooklyn Dodgers, in the New York City of my youth. In 1956 -- the same year as President Dwight D. Eisenhower signed into law the Interstate Highway Bill (indeed, he also threw out the first ball in the series) -- New York was the biggest city in the world by population. Many things have changed on this planet since 1956, but perhaps nothing so substantial -- at least in urban life -- as the phenomenon of the mega-city. These enormous metropolitan regions, growing in both the developed and the developing worlds, have relegated New York to a more modest place among the large cities of the world and present us with some of the most challenging transportation and environment issues we currently face.

The mega-city trend has been most evident in developing countries, where cities like Bangkok, Thailand; Djakarta, Indonesia; São Paulo, Brazil; Mexico City, Mexico; Beijing, China; Bombay, India; and others have achieved gigantic size in a relatively short number of years. In these developing-country mega-cities, oftentimes more affluent segments of the population live close to the center city, while less affluent citizens live on the outskirts (the opposite of patterns in the U.S.), causing substantial problems, both in mobility and in accessibility. The growth of Mexico City, arguably the largest city in the world, with a population of 20 million, is fueled by a search for a better life by Mexicans living in agricultural regions, often settling in "illegal" settlements at the outer fringes of this burgeoning metropolitan region. Some additional sprawl is caused by "corona" cities, far from downtown, with more affluent residents.

---

[1] Reprinted with permission of ITS America. Sussman, Joseph M., "Mega-cities in Developing Countries -- a Major ITS Market for the Future", "Thoughts on ITS" Column, *ITS Quarterly*, ITS America, Washington, DC, Fall 2000.

The Transportation/Land-Use/Environment connection is of particular importance in these mega-cities. The substantially uncontrolled land-use patterns in Mexico City cause growth in transportation needs (VMTs) and in environmental impacts.

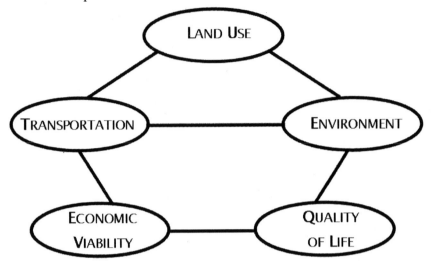

Viable transportation is important to maintain economic activity and a growth in living standards. At the same time, transportation often causes the lion's share of the air-quality problems in these cities.

How can ITS help in the management and improvement of quality of life in the developing-country mega-city context?

Rates of automobile ownership are substantially lower in such cities, as compared with the developed world, and people are more dependent on the public transportation system, which is characterized by a larger set of options than one sees in the developed world. Of special interest are "informal" modes -- "jitneys" -- called "colectivos" in Mexico City, "públicos" in San Juan, Puerto Rico, and by various other names around the world. This informal transportation is characterized by a very large number of relatively small vehicles (capacity less than 20) running in a semi-regulated and often non-fixed-route manner around the city. While uncoordinated, these jitneys collectively provide a relatively high level-of-service because of their high frequency. In Mexico City, they have higher fares than conventional buses or the metro system, but their superior level-of-service (as compared with buses) makes them the mode of choice for many public transportation users. In fact, in Mexico City, the colectivos have put conventional buses all but out of business. At the same time, their mode of operation (e.g., stopping in the middle of the street to pick up and drop off passengers, racing from block to block with the competing buses and other jitneys to get to the passengers first) can cause chaos in the streets of an already-congested city.

The question: How to take advantage of the positive aspects of jitney service and ameliorate the negative side effects?

The answer: Developing a coordinated jitney organizational structure will be critical, and ITS technology can be a mechanism for achieving that coordination. The small vehicle size, high frequency and flexible nature of the services offered, when coupled with ITS technology, can be a partial answer to mobility needs in developing-country mega-cities.

Indeed, the idea of using ITS to coordinate intermodal public transportation trips, taken by passengers who might use jitneys, buses and metros to get from origin to destination, can be an important service innovation in mega-cities.

In Mexico City, while rates of automobile ownership are low, they are growing rapidly. As has been observed again and again, when per capita income begins to rise above certain thresholds, automobile ownership rises dramatically. And often there is not the physical infrastructure to serve all these additional cars. Again, this is a classic application for ITS. The use of our ITS technologies for network control, traveler information and road pricing can help here, particularly where there is a paucity of conventional infrastructure.

Finally, those who have worked in the developing-country context know that data to support effective transportation (and environmental) system analyses is often sadly lacking. Once again, ITS can come to the fore. The most recently-added service to the national ITS architecture has been archiving of transportation data. Such databases in the developing-country context would be of extraordinary value in allowing transportation, planning and environmental professionals to devise effective strategies for future systems.

The management of the developing-country mega-city is one of our most important urban transportation and environmental issues. ITS can help to deal with the problems those mega-cities face. The Interstate Highway System, at the time of the last "Subway Series", dealt visibly and effectively with the issue of surface intercity mobility. Now, ITS, 44 years later, less visibly and more subtly, but no less importantly, helps us deal with the developing-country mega-city, a critical international concern and, importantly, a *major ITS market* for the future.

SECTION V. WHERE WE ARE TODAY IN ITS AND ISSUES FOR THE FUTURE

"Good ideas are not adopted automatically. They must be driven into practice with courageous patience."

-- Admiral Hyman Rickover

The final section captures the state-of-the-art in ITS in the United States and the challenges for the future. The first article is derived from a DOT report entitled, "What Have We Learned About Intelligent Transportation Systems?" I was responsible for writing the synthesis of the state-of-the-art reports written by seven industry experts representing various segments of the ITS program. All these reports are contained in "What Have We Learned About Intelligent Transportation Systems?" My synthesis both summarized their findings and tried to distill the overarching, cross-cutting conclusions about ITS. Then, I point the way to the future and what we in the ITS community must achieve if we are to see this concept continue to advance and thrive and have positive impact on the world of transportation and related social systems.

The second article discusses twenty important *transitions* in the transportation world, many enabled or even forced by ITS.

The final article takes a retrospective look at the 1991/92 ITS Strategic Plan ("A Strategic Plan for IVHS in the United States"), which has been characterized as an early seminal document for the ITS field. My article contrasts the collective view of ITS back then with what we now know about ITS and how the technology has been effectively deployed (or not).

For additional articles relating to these themes, also see Section IV-6.

V. 1. INTELLIGENT TRANSPORTATION SYSTEMS AT THE TURNING POINT: PREPARING FOR INTEGRATED, REGIONAL, AND MARKET-DRIVEN DEPLOYMENT[1]

Intelligent transportation systems (ITS) apply technologies in communications, control, electronics, and computer hardware and software to improve surface transportation system performance.

1. INTRODUCTION

This simple definition points to a substantial change in surface transportation. The increased social, political, and economic difficulty of expanding transportation capacity through conventional infrastructure-building has motivated the development of ITS. ITS represents an effort to harness the capabilities of advanced technologies to improve transportation on many levels -- to reduce congestion, enhance safety, mitigate the environmental impacts of transportation systems, enhance energy performance, and increase productivity.

The U.S. National ITS Program is more than a decade old. In December 1999, the ITS Joint Program Office of the U.S. Department of Transportation initiated a project, What Have We Learned About ITS?, with a series of presentations in Washington, DC. Industry experts responded to the following questions about ITS technologies and applications:

- What ITS technology applications have been successful and why?
- What ITS technology applications have not been successful and why?
- For which ITS technologies is "the jury still out"?
- What institutional issues arose in ITS deployments and how were they overcome?
- What next steps are needed?

---

[1] Reprinted with permission of the Transportation Research Board. Sussman, Joseph M., in TR News, January-February 2002, Number 218, Transportation Research Board, National Research Council, Washington, DC, pp. 10-17.

In April 2000, in conjunction with the Institute of Transportation Engineers 2000 International Conference in Irvine, California, the initial results were presented to a broader community to validate or debunk -- and, if possible, to document a national consensus. The following summarizes some of the key concepts presented in papers by experts in the seven ITS areas studied (*1*).[2]

2. FREEWAY, INCIDENT, AND EMERGENCY MANAGEMENT, AND ELECTRONIC TOLL COLLECTION (ETC)[3]

This area comprises several different, but related, technologies: Transportation management centers, ramp metering, dynamic message signs, roadside infrastructure, and dynamic lane and speed control. ETC is one of the fundamental and earliest-deployed ITS technologies -- it is the most common example of the electronic link between vehicle and infrastructure that characterizes ITS.

Freeways, or limited-access highways, are another major and early ITS application area. Incident management is important in reducing nonrecurring congestion on freeways. Emergency management predates ITS, but is enhanced through ITS technologies.

Several systems have gained deployment; however, more remains to be accomplished. An important technical advance would upgrade the systems to predict congestion from current traffic patterns and expectations, replacing responsive systems. Institutions need to establish operations budgets for these systems and to attract high-quality technical staff for deployment and operations support.

---

[2] "What Have We Learned About Intelligent Transportation Systems?", U.S. Department of Transportation, Washington, DC, 2000.

[3] Pearce, Vincent (Booz-Allen & Hamilton, now with FHWA), "What Have We Learned About Freeway, Incident, and Emergency Management and Electronic Toll Collection?", Chapter 2 in "What Have We Learned About Intelligent Transportation Systems?", U.S. Department of Transportation, Washington, DC, 2000.

3.   ARTERIAL MANAGEMENT[4]

The management of arterials -- high-capacity roadways controlled by traffic signals, with access via cross-streets and often from abutting driveways -- predates ITS, with early deployments in the 1960s. Nonetheless, adaptive control strategies for arterials, making real-time adjustments to traffic signals based on conditions such as queues, are not in widespread use.

The reasons for this lag include costs as well as concerns that adaptive traffic control algorithms do not perform well. When traffic volumes are heavy, state-of-the-art algorithms appear to break down -- although vendors claim otherwise. The complexity of the system also requires additional training for personnel.

Traveler information systems for arterials are not yet widely deployed, although studies suggest safety benefits and reductions in delays. Cellular phones, traffic probes with cellular phone geolocation, and implementation of the national three-digit traveler information number (511) may stimulate deployment. Integrating traffic management technologies -- such as emergency vehicle management, transit management, and freeway management -- with arterial management may be an important next step.

4.   TRAVELER INFORMATION SYSTEMS[5]

Traveler information is one of the core concepts of ITS. Travelers value easy and timely accessibility to high-quality information, high-quality user interfaces, and low prices -- preferably free. Consumer demand for traveler information is a function of

- The amount of congestion on the regional transportation network,
- The network's characteristics,
- The quality of the information and the user interface,
- The characteristics of individual trips, and
- The characteristics of drivers and transit users.

Many kinds of traveler information systems are in use. Although people value high-quality travel information, they are not necessarily willing to pay for it, since free information is available, such as radio reports. Whether

---

[4] Hicks, Brandy and Mark Carter (SAIC), "What Have We Learned About ITS? Arterial Management", Chapter 3 in "What Have We Learned About Intelligent Transportation Systems?", U.S. Department of Transportation, Washington, DC, 2000.

[5] Lappin, Jane (EG&G Technical Services/John A. Volpe National Transportation Systems Center), "What Have We Learned About Advanced Traveler Information Systems and Customer Satisfaction?", Chapter 4 in "What Have We Learned About Intelligent Transportation Systems?", U.S. Department of Transportation, Washington, DC, 2000.

traveler information systems can be viable as a stand-alone commercial enterprise is unclear; transportation information probably will be packaged with other information services via the Internet.

Traveler information systems make clear that ITS operates within the environment of people's expectations for information. Timeliness and quality of information are continually increasing for many non-ITS applications, such as the Internet, and providers of traveler information need to be aware of changing expectations.

The integration of traveler information with network management or transportation management systems, such as freeway and arterial management, has not occurred for the most part. Network management and traveler information systems would benefit from substantial integration, as would the customers -- travelers and freight carriers.

5.   ADVANCED PUBLIC TRANSPORTATION SYSTEMS[6]

Transit has difficulty attracting market share for the following reasons:
- Land-use patterns incompatible with transit use;
- Lack of high-quality service, with long and unreliable travel times;
- Lack of comfort;
- Security concerns; and
- Incompatibility with the way people currently travel -- for example, by trip-chaining.

ITS transit technologies -- including automatic vehicle location, passenger information systems, traffic signal priority, and electronic fare payment -- can improve transit productivity, quality of service, and real-time information. However, deployment of ITS to upgrade transit has been modest, stymied by
- A lack of funding for ITS equipment;
- Difficulties in integrating ITS technologies into conventional transit operations; and
- The lack of human resources to support and deploy the technologies.

As people with ITS expertise join transit agencies, there will be a steady but slow increase in the use of ITS technologies for transit management. But

---

[6] Casey, Robert (John A. Volpe National Transportation Systems Center), "What Have We Learned About Advanced Public Transportation Systems?", Chapter 5 in "What Have We Learned About Intelligent Transportation Systems?", U.S. Department of Transportation, Washington, DC, 2000.

training is needed, and a chronically capital-poor industry must overcome inertia to deploy these technologies.

Integrating transit services with other ITS services promises major intermodal benefits; the integration of highway and transit, multiprovider services, and intermodal transfers may be feasible in the near term. But the transit industry should provide critical, high-quality service in urban areas and can support environmentally-related programs -- ITS may be the mechanism to boost and reinvent the industry.

6.   COMMERCIAL VEHICLE OPERATIONS[7]

Through commercial vehicle operations (CVO), states ensure safety and enforce regulations related to truck operations on highways; the public-sector components of CVO are the main focus. Commercial vehicle information systems and networks (CVISN) deal with roadway operations, including safety information exchange and electronic screening, as well as back-office applications like electronic credentialing.

CVISN has experienced some successes. In most programs, participation by carriers is voluntary; requiring truckers to use transponders may be difficult -- universal deployment is a challenging task. Another problem for deployment is consistency from state to state. Because trucking is a regional or even national business, the interface between the trucking industry and the states must be consistent for widespread deployment. Although each state has its own requirements based on the operating environment, interstate interoperability is necessary through expanded partnerships among states and between the federal government and states.

The CVISN program has raised some public- and private-sector tensions. For example, truckers endorse the technology that allows weigh-station bypasses for previously checked vehicles -- the information is relayed from the adjoining station or even from another state. Yet because of competition, truckers are concerned about another application of the same system -- for tax collection -- questioning its equitability and the privacy of origin-destination data. Public-private partnerships need to develop both applications to capture the benefits effectively.

---

[7] Orban, John (Battelle), "What Have We Learned About ITS for Commercial Vehicle Operations? Status, Challenges, and Benefits of CVISN Level 1 Deployment", Chapter 6 in "What Have We Learned About Intelligent Transportation Systems?", U.S. Department of Transportation, Washington, DC, 2000.

7. CROSSCUTTING TECHNICAL AND PROGRAMMATIC ISSUES[8]

Advanced technology is at the heart of ITS, which means dealing with changing technologies while relating to the need for standards. Rapid obsolescence is a problem, but technology issues are not a substantial barrier to ITS deployment; costs, however, can be a barrier. Most technologies perform -- but are they priced within the budget of the deploying organizations, and are the prices consistent with the benefits?

Surveillance and communication are two core ITS technologies. Surveillance technologies have experienced successes with cellular phones for reporting and videos for verifying incidents, but cellular phone geo-location for traffic probes is still a question. The lack of traffic-flow sensors in many areas and on some roadway types inhibits the growth of traveler information and the improvement of transportation management systems.

Communications technologies have experienced success with the Internet for pre-trip traveler information and credentials administration in CVO. The growth rate in the use of the Internet and also emerging technologies like the wireless Internet and automated information exchange may portend increased use of ITS applications.[9]

8. CROSSCUTTING INSTITUTIONAL ISSUES[10]

The key barriers to ITS deployment are institutional, involving issues such as the awareness and perception of ITS, long-range operations and management, regional deployment, human resources, partnering, ownership and use of resources, procurement, intellectual property, privacy, and liability.

Public awareness and political appreciation that ITS can help deal with issues such as safety and quality of life are central to successful deployment. Building a regional perspective on deployment through public-private

---

[8] McGurrin, Michael (Mitretek Systems), "What Have We Learned About Crosscutting Technical and Programmatic Issues?", Chapter 7 in "What Have We Learned About Intelligent Transportation Systems?", U.S. Department of Transportation, Washington, DC, 2000.

[9] "A Survey of Government on the Internet: The Next Revolution", *The Economist*, June 14, 2000.

[10] DeBlasio, Allan J. (John A. Volpe National Transportation Systems Center), "What Have We Learned About Crosscutting Institutional Issues?", Chapter 8 in "What Have We Learned About Intelligent Transportation Systems?", U.S. Department of Transportation, Washington, DC, 2000.

partnerships is important. Planning for sustained funding for long-term operations also is critical. Procurement is an institutional concern, and public-sector agencies are not accustomed to procuring high-technology components that may involve questions of intellectual property.

ITS deployment requires a cultural change for transportation organizations that traditionally have focused on conventional infrastructure, not on operations. This cultural change is a continuing, ongoing, arduous process that must be undertaken if ITS is to be deployed successfully.

9.   ASSESSING ITS

An assessment of ITS should consider the three dimensions that characterize transportation: technology, systems, and institutions (2):
- Technology includes infrastructure, vehicles, and the hardware and software that make them function.
- Systems deal with the performance of holistic sets of components -- for example, a regional transportation network.
- Institutions refer to organizations and interorganizational relationships that support the development and deployment of transportation programs.

9.1   Technology

Four technologies are central to most ITS applications:
- Sensing -- registering the position and velocity of vehicles on the infrastructure;
- Communicating -- from vehicle to vehicle, between vehicle and infrastructure, and between infrastructure and centralized transportation operations and management centers;
- Computing -- processing the data collected and communicated during transportation operations; and
- Algorithms -- computerized methods for operating transportation systems.

In most cases, off-the-shelf technology can support ITS functions. The important questions about technology quality concern algorithms -- for example, the efficacy of software to perform adaptive traffic signal control. Also, the quality of the information collected may be a technical issue in some applications.

Public agencies may see the technology as too costly for deployment, operations, and maintenance, particularly if the benefits to be gained are not commensurate. In some cases, technology falters because it is not easy to use -- intuitive user interfaces are essential.

9.2   Systems

The integration of ITS components is the critical need at the systems level. Many ITS deployments are stand-alone applications, such at ETC. IT is often cost-effective in the short run to deploy an application without worrying about the interfaces and platforms required for an integrated system. Decisionmakers often have opted for stand-alone applications -- a reasonable approach for the first generation of ITS deployment.

However, the next steps require system integration for efficiency and effectiveness -- for example, integrating services for arterials, freeways, and public transit, then integrating incident management, emergency management, traveler information, and intermodal services. Integration adds complexity, but also provides economies of scale in system deployment and improvements in overall system effectiveness, resulting in better freight and traveler services.

Another aspect of system integration is interoperability -- ensuring that ITS components can function together. Possibly the best example is the interoperability of ETC hardware and software in vehicles and on the infrastructure. To achieve interoperability, the design of electronic linkages among vehicles and infrastructure must employ system architecture principles and open standards.

The public wants transponders that will work with ETC systems across the country or even regionally. The technology should operate not only on a broad geographic scale, but also locally for public transportation and parking applications.

Systems that should work on a national scale, such as CVO, must achieve interoperability among components. There are institutional barriers to interoperability -- for example, the differences among jurisdictions -- although widespread deployment is ultimately in the interest of all.[11]

Integration is needed between advanced transportation management systems (ATMS) and advanced traveler information systems (ATIS); the two technologies have developed largely independently. ATMS provides for operations of networks and ATIS for pre-trip and in-vehicle information for individual travelers. ATMS can collect and process a variety of network status data and can estimate future demand to provide travelers with

---

[11] Orban, John (Battelle), "Commercial Vehicle Operations", Chapter 6 in "What Have We Learned About Intelligent Transportation Systems?", U.S. Department of Transportation, Washington, DC, 2000. Orban contrasts technical interoperability, operations interoperability, and business model interoperability in the context of CVO and CVISN.

dynamic rout guidance via ATIS services. With integration, ATMS-derived operating strategies for the network -- which account for customer response to ATIS-provided advice -- can lead to better network performance and better individual routes.

9.3   Institutions

Technical integration is vital, but institutional integration will be equally important for the future of ITS, including the integration of public- and private-sector perspectives on ITS, as well as the integrated operations of various public-sector organizations.

The major barriers to ITS deployment are institutional. Looking at transportation from an intermodal, systemic point of view requires a shift in institutional focus. Dealing with intra- and interjurisdictional questions, budgets, and regional perspectives on transportation systems; shifting institutional attention to operations instead of construction and maintenance; and training, retaining, and compensating qualified staff are institutional barriers to deployment. Developing strategies to overcome these institutional barriers is the single most important activity to ensure successful ITS deployment and implementation.

10.   ITS AND OPERATIONS

In recent years, transportation operations -- as opposed to construction and maintenance of infrastructure -- have become a primary focus. ITS deals with the technology-enhanced operations of complex transportation systems. The ITS community has argued that focusing on operations through advanced technology is cost-effective, considering the social, political, and economic barriers to conventional infrastructure, particularly in urban areas. ITS can avoid the high upfront costs of conventional infrastructure through more modest investments in electronic infrastructure, followed by a focus on effectively operating the infrastructure and the transportation network at large.

Although ITS can provide less expensive solutions, there are upfront infrastructure costs and additional expenses for operating and maintaining hardware and software. Training staff to support operations requires resources. Spending for ITS differs from spending for conventional infrastructure, requiring less up front but more investment in the following "out" years. Therefore, planning for operations requires a long-term perspective by transportation agencies and politicians.

Operations should be institutionalized within transportation agencies. To maintain system effectiveness and efficiency, budgets for operations must be stable and cannot be subject to yearly fluctuation and negotiation. Human resources needs also must be considered.

To justify ITS capital as well as continuing costs, it is helpful to consider life-cycle costs -- the costs and benefits that accrue over the long term are the important metric. But organizations must recognize that a lack of follow-through will cause out-year benefits to disappear if unmaintained ITS infrastructure deteriorates and if the algorithms for traffic management are not recalibrated.

10.1   Mainstreaming

Mainstreaming has several definitions in the ITS context. To some, mainstreaming means integrating ITS components into conventional projects. Two examples of projects that include conventional infrastructure and ITS technologies and applications are the Central Artery-Ted Williams Tunnel project in Boston, Massachusetts, and the redesign of the Woodrow Wilson Bridge on I-95, connecting Maryland and Virginia. In such projects, the ITS component typically is a fraction of the total project cost.

Nonetheless, ITS technologies and applications can come under close political scrutiny disproportionate to their financial impact. For example, in the Woodrow Wilson Bridge project several ITS elements were considered for elimination (3).[12]

Another approach to mainstreaming suggests that ITS projects should not be protected by specially allocated funds, but should compete for funds with all other transportation projects. The advantage is that ITS would compete for a larger pool of money; the disadvantage is that ITS might not compete successfully. Those responsible for spending public monies have favored conventional projects for transportation infrastructure. Convincing decision-makers that the funds are better spent on ITS applications may be difficult.

This issue is also linked to human resource development. Professionals cannot be expected to select ITS unless they are knowledgeable about it; education of the professional transportation cadre is essential for mainstreaming. The National ITS Program also must demonstrate that the benefits of ITS deployments are consistent with the costs.

Protected funds that can be spend only on ITS applications may be a good transitional strategy as professional education continues and the

---

[12] John Collins, then President of ITS America, likened the decommitting of ITS technologies from the Woodrow Wilson Bridge to "constructing a house and deciding to save money by not buying light bulbs".

benefits become more clear; but in the long run, mainstreaming ITS via competitive proposals will be advantageous.

10.2 Human Resources

The deployment of the new ITS technologies and applications requires personnel -- skilled, knowledgeable specialists, as well as generalists with policy and management skills who can incorporate advanced thinking about transportation technologies and services into systems (4).

Several organizations have established programs for human resource development -- for example, the Federal Highway Administration's (FHWA's) Professional Capacity Building program and CITE (Consortium for Intelligent Transportation Education), housed at the University of Maryland. These programs, along with graduate transportation programs incorporating ITS-related changes, can prepare talented and skilled people for the industry.

However, institutional changes in transportation organizations are needed to engage and retain personnel with high-technology skills, who often can demand higher salaries than public-sector transportation organizations can provide. Cultural change, along with appropriate rewards for operations staff, will be necessary in organizations that have favored conventional infrastructure construction and maintenance.

Public-sector organizations may have to contract for outside staff to perform some high-technology functions. Contracting with private-sector organizations to handle various ITS functions is another option. In the short run, these options may be helpful; in the long run, however, developing technical and policy skills within the public agency has advantages for strategic decisionmaking.

11. ITS OPPORTUNITIES

11.1 Regional Approaches

ITS provides an opportunity to manage transportation at the scale of the metropolitan-based region. Along with state or multistate geographic areas, metropolitan-based regions -- the basic geographic unit for economic competition and growth (5) and for environmental issues -- can manage transportation effectively through ITS.

A few regions have made progress, although none yet has translated ITS technologies into a complete, regionally-scaled capability. Thinking through the organizational changes to allow some autonomy for subregional units, but also system management at the regional scale, is a priority (6). The strategic vision is for ITS to integrate transportation, communications, and intermodalism on a regional scale (7). Multistate regions with traffic coordination over large geographic areas, such as the mountain states -- and also corridors, such as I-95, present ITS opportunities.

11.2 Surface Transportation Markets

Surface transportation should be thought of as a market of individual customers with ever-rising and differing expectations. Modern markets provide choices. People demand choices in level of service and often are willing to pay for superior service; surface transportation customers increasingly will demand this service differentiation. Although a market framework for publicly-provided services is not without controversy, surface transportation operators can no longer think in terms of "one size fits all".

An early example of this market concept in highway transportation is the high-occupancy toll lane, which uses ITS technologies to allow single-occupant vehicles on a high-occupancy vehicle lane for a toll. Other market opportunities building on ITS will emerge, as researchers and policymakers consider how surface transportation should operate in relation to markets -- philosophically and conceptually -- in an ITS environment.

A market or customer focus plus the constraints on building conventional infrastructure require an emphasis on operations enabled by ITS technology. Technological change and an emphasis on operations, in turn, entail changes in transportation organizations. The institutional changes for operations involve different funding arrangements as well as sharing information and responsibility on a regional scale (Figure 1).

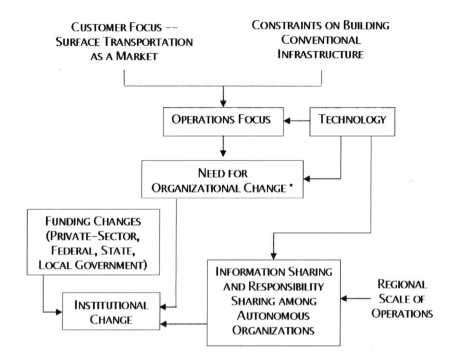

FIGURE 1. *CHANGES IN A REGIONAL INTELLIGENT TRANSPORTATION SYSTEMS (ITS) ENVIRONMENT*

11.3   Introducing Change

ITS presents a turning point in surface transportation, similar to the introduction of air traffic control systems into air transportation. In scale, ITS resembles the Federal-Aid Highway Program, which forged a new relationship between the federal and state governments as the idea of a national highway system took shape during the second decade of the 20th century.

The electronic linkage between vehicle and infrastructure via ITS has profound implications for surface transportation. But so far the changes have been incremental; the real impact has yet to be felt. Integrated, regional systems are examples of the changes to come.

The functional change that ITS introduces must go beyond institutional changes in transportation organizations to cultural changes -- reflecting the importance of operations, new technology, and market-based forces, especially in the highway sector. Achieving these cultural changes will take leadership, education, and training.

ITS offers an opportunity for the transportation profession to evolve to a more sophisticated level. Advanced technologies, system-thinking about transportation services, and expanded possibilities for policy initiatives in technology-enabled transportation create vital professional opportunities -- which the educational sector must recognize and develop.

12.   GREAT EXPECTATIONS

What have we learned about ITS? Much has been achieved by choosing clear-cut, sure winners -- an appropriate strategy for the first generation of any technology. However, successful deployment requires focusing on integrated, regional, and market-driven systems.

ITS can be a critical component of surface transportation. The public's expectations are changing in the age of the Internet. People are using sophisticated information technology and telecommunications equipment every day -- the expectation is for accessible information from multiple sources at the click of a mouse or television switch.

ITS is the transportation community's opportunity to be part of this revolution and to advance transportation and the profession. Success will be predicated on extensive deployment and on integrated, regional systems. For now, ITS is on the right track, but more must be achieved, as more will be expected.

References

1. "What Have We Learned About Intelligent Transportation Systems?", U.S. Department of Transportation, Washington, DC, 2000.
2. Sussman, J. M., *Introduction to Transportation Systems*, Artech House, Boston, Mass., and London, 2000.
3. *The Washington Post*, June 29, 2000, p. A-15.
4. Sussman, J. M., "Educating the 'New Transportation Professional'", *ITS Quarterly*, ITS America, Washington, DC, Summer 1995.
5. Porter, M., *On Competition*. Harvard Business School Press, Boston, Mass., 1998.
6. Hardy, C., "Are We All Federalists Now?" In *Beyond Certainty: The Changing Worlds of Organizations*, Harvard Business School Press, Boston, Mass, 1996.
7. Sussman, J. M., ITS Deployment and the 'Competitive Region', *ITS Quarterly*, ITS America, Washington, DC, Spring 1996.

V. 2. TRANSITIONS IN THE WORLD OF TRANSPORTATION: A SYSTEMS VIEW[1]

In 1995, I wrote "Educating the 'New Transportation Professional'", which described a T-shaped professional, with breadth represented by capabilities in technology, systems and institutions, and some in-depth specialization in one of those areas.[2]

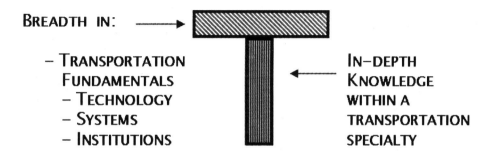

BREADTH IN:

- TRANSPORTATION
  FUNDAMENTALS
- TECHNOLOGY
- SYSTEMS
- INSTITUTIONS

IN-DEPTH
KNOWLEDGE
WITHIN A
TRANSPORTATION
SPECIALTY

The horizontal bar reflects deep professional expertise; the breadth enables the new transportation professional to perform as part of a team with other specialists.

This "new transportation professional" must deal with the important transitions which characterize the field. This article identifies some of these transitions in the transportation field, with an eye toward understanding their strategic implications, and what these changes imply for the "new transportation professional".

---

[1] Reprinted with permission of the Eno Foundation. Sussman, Joseph M., "Transitions in the World of Transportation: A Systems View", *Transportation Quarterly*, Vol. 56, No. 1, Winter 2002, Eno Transportation Foundation, Washington, DC, 2002.

[2] Sussman, Joseph M., "Educating the 'New Transportation Professional'", *ITS Quarterly*, ITS America, Washington, DC, Summer 1995. N.B. This article appears in Section III of this volume.

TRANSITIONS

1.   **From**   **To**

     Capital Planning   ————————▶   **Management
                                                            and Operations Focus**

For many decades the focus of transportation planning has been large-scale capital facilities. In the U.S., the Interstate system, the linchpin of the transportation program for four decades, is an exemplar of this approach. Provision of large-scale highway facilities connecting our cities and providing transportation in and around them has dominated our thinking.

Now it is clear that we must shift away from the large-scale infrastructure planning focus, especially in urban areas. Building more urban capital facilities is quite difficult from a cost, as well as political and social, standpoint. The emphasis now must be on more effectively *managing* and *operating* existing capital facilities (although, of course, some building will continue). Technologies such as Intelligent Transportation Systems (ITS) are at the heart of this approach. These technologies can allow us to manage and operate capital facilities more effectively, squeezing more capacity out of them, in the highway and public transportation domains. Similarly, in air transportation, technologies to expand airport capacity without new runways have been emphasized.

This focus on operations will cause an important cultural shift in many of our transportation institutions, particularly in the public sector. An operations "mindset" must be introduced.[3] These organizations must change their focus so that the importance of transportation *operations* -- rather than building capital facilities -- is reflected in their organizational structure and hierarchy and in providing rewards and incentives so that top-quality transportation people will focus their attention on operations.

2.   **From**   **To**

     Long Timeframes   ————————▶   **Real-time Control**

Related to (1) above, the time-scale at which we consider the transportation system has changed. Strategic plans, with long timeframes for creating conventional infrastructure, formerly our primary purpose, have been augmented with operating systems in *real-time*. Technology has permitted us to manage and control transportation networks on literally a second-to-second basis. This is complementary with infrastructure planning

---

[3] Sussman, Joseph M., "Transportation Operations: An Organizational and Institutional Perspective", Operations Summit, http://www.ite.org/NationalSummit/index.htm, Columbia, MD, October 2001. N.B. This article appears in Section II of this volume.

in that real-time management and control allows us to utilize conventional capacity more effectively. With the advent of real-time capability, we need algorithms in areas such as dynamic traffic assignment for surface networks, which will utilize this information effectively.

There is much to be done. Our transportation organization will need individuals with the technical strengths to create systems that can effectively respond in real time; the technologies to do so are just enablers of more efficacious real-time system operation.

3.	**From**		**To**
	Urban Scale		**Regional Scale**
	Planning and	⎯⎯⎯⎯⎯⎯⎯⎯→	**Planning and**
	Operations		**Operations**

For many years, operations have been managed at the scale of the urban area; for example, traffic light coordination has been with us for decades. But we are embarking now on an era in which the metropolitan-based region is pre-eminent. Michael Porter and others have made a compelling case that the fundamental unit of competitiveness is the metropolitan-based region (as opposed to the nation-state).[4] Clearly the effectiveness of transportation services in a region, and the availability of good interregional connections, will contribute to the competitiveness of that region on the national and the international scales.

Further, environmental specialists suggest that the management of environmental problems needs to take place on a metropolitan-based regional scale and not only the urban scale. Decisions made on an urban scale to mitigate environmental problems will be doomed to failure unless a metropolitan-based planning structure is adopted.

So regions are a critical concept in our modern era and through technologies (like ITS) we now have the capability of managing integrated regionally-scaled transportation systems.[5] With competitive and environmental issues being considered at a metropolitan-based regional scale, we have the capability of managing *transportation* at that regional scale as well.

Again, we must not overlook the considerable institutional issues that we will face in establishing such regionally-scaled transportation management structures. Self-interest will often trump broader regional interests in designing and operating transportation services. But we *can* say that technology is *not* the barrier to accomplishing regional transportation

---

[4] Porter, Michael, *On Competition*, Harvard Business School Press, Boston, MA, 1998.

[5] Conklin, Christopher and Joseph M. Sussman, "Regional Architectures, Regional Strategic Transportation Planning and Organizational Strategies", ITS America 2000 Annual Meeting, Boston, MA, May 2000.

management, which many would argue is an inevitability and, indeed, a critical *goal* for the future.

4.   **From**   **To**

                                              Emphasis on Accessibility
Emphasis on Mobility   ──────────▶   **(the Transportation/**
                                              Land-Use Connection)

In earlier decades, the emphasis in transportation was providing *mobility*, the ability to go from anywhere to anywhere else at any time, in an uncongested fashion. While impossible to achieve, the mobility concept drove transportation investment.

Now, we recognize the importance of *accessibility* as a complementary concept. As suggested by Ming Zhang[6],

> "Put in a succinct way, mobility refers to the ability to move between different places whereas accessibility refers to the ability to reach destination opportunities under the constraints of time or cost."

The interaction of land-use and transportation creates *accessibility* to *meaningful destinations* for economic and social purposes. This accessibility results not only from mobility on the transportation network but land-use patterns as well (i.e., origins and destinations). And so land use is increasingly on the agenda; consider the current concern with suburban sprawl and redensification of our cities.

Further, the concept of accessibility can be extended to the idea of using telecommunications technology to substitute for travel, as in telecommuting or remote-conferencing and remote-teaching. Thinking more broadly about communications as a "transportation" mode and transportation/communication complementarity is of relevance here.

---

[6] Zhang, Ming, "Job Accessibility in the San Juan Metropolitan Region (SJMR): Maximizing the Benefits of Tren Urbano", Doctoral Thesis, Massachusetts Institute of Technology, Cambridge, MA, May 1999.

5.   <u>**From**</u>   <u>**To**</u>

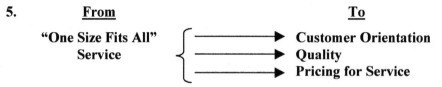

"One Size Fits All"   **Customer Orientation**
Service   **Quality**
                                                       Pricing for Service

This is the era of the *customer*. It is a business and (in some instances) public-sector axiom that organizations must focus on the needs of their customers and the attraction of new ones. Further, customers are becoming more demanding, wanting *immediate*, *perfect* and *cheap* service or artifacts in many dimensions of their lives.

This suggests the need for a major transition in the transportation industry, and the surface transportation industry in particular. Our highways and our public transportation systems have traditionally been designed on a "one size fits all" basis. We provide a non-differentiated service and expect everyone to use it in the same way. We make no attempt to differentiate among customers who are willing to pay a premium price for a premium service, and other customers less willing to do so. Indeed, the very word *customer* has been foreign to many transportation organizations.

It is critical that we adopt a customer orientation supported by the management and operations focus discussed under (1). We must provide quality, and price transportation in such a way that people willing to pay for premium service can pay for it *and* then actually receive it.

Perhaps the simplest example of this in highways is the high-occupancy toll (HOT) lane. This is an offshoot of the high-occupancy vehicle (HOV) lane, in which one is permitted to use this (presumably less-congested) lane if one has a certain number of people in the car. The HOT lane extends this to allow vehicles with, for example, only a driver to use that lane for a premium price (that is, a toll) to purchase a faster and more reliable travel time. Examples in public transportation can be developed as well and, of course, transportation demand management is part of the equation as we seek to provide flexible, high-quality services for the customers of our transportation system. We can operationally perform in this manner; the question again will be, "Will our institutions and cultures change to allow this to take place in an effective manner?"

6.   <u>**From**</u>   <u>**To**</u>

Allocate Capacity   **Allocate Capacity**
by Queuing   **by Pricing**

A related idea deals with the transition from allocating capacity -- particularly highway capacity -- by *queuing* to allocating capacity by *pricing* for road services. Our highway system has long been run in a "Soviet-style" manner, with capacity being allocated by people simply queuing for service, as in the butcher shops of the old USSR. When you got to the front of the line, if no more meat was available, that was your problem; the butcher shop allocated a scarce resource (meat) via queuing. Our highway system has operated in a comparable way, particularly during rush hours, with its limited capacity allocated to those willing to wait.

There is a better way and one that has been in use by many systems that are infrastructure-intensive. The reason that movie theaters charge $8 on Saturday night and $2 on Tuesday afternoons recognizes the substantial unused capacity at the latter time and represents an attempt by the movie operator to "shave the peak" on Saturday night by enticing some people to his/her almost empty theater on Tuesday afternoon. The purchase of electric power has long operated in that same way, with on- and off-peak pricing policies.

In the 1950s, William Vickery of Columbia, a Nobel laureate in economics, recognized that this concept could be applied to transportation systems through the idea of congestion pricing: charging people for road use as a function of time of day or congestion levels, and then presumably giving them the incentive to move off the peak and onto the shoulders of the traditional diurnal distribution. While theoretically sound, the technology to deploy this idea has not been available until recently. Now, as an extension of electronic toll collection, we can vary the electronic tolls as a function of time of day or degree of congestion, and have the ability, in principle, to attract people off the peak onto the "shoulders".

This idea can be extended to pricing other previously unpriced externalities, such as environmental impacts. For example, high "environmental" tolls on "bad air" days could be implemented.

So pricing for transportation services, a "first cousin" to the customer orientation described earlier, is a critical transition that we in the transportation world face. This is simply a natural extension of the *demand management* ideas in use for some years, as opposed to purely supply-side options.

7.	**From**		**To**
	Aggregate	→	Disaggregate
	Methods for		Methods for
	Demand Prediction		Demand Prediction

Prediction of demand for both traveler and freight transportation is a difficult conceptual and practical area of transportation systems analysis. While prediction of transportation demand is needed to make good investments in transportation infrastructure and perform effective operations management of the resulting systems, reliable predictions have been difficult to come by.

However, we have seen substantial progress over the years. Transportation demand was formerly predicted through what we call aggregate methods, where the general characteristics of a geographic zone (economic, social and so forth) were used to predict its trip-generating power as well as its trip-attraction power. Often the way the trips were "distributed" from a generating area to an attracting area was through physically appealing but nonetheless flawed methods, such as the Gravity Model (where the attraction between two zones is proportional to the inverse square of the distance -- or perhaps travel time -- between them, à la Sir Isaac Newton).

In recent years *disaggregate* approaches have been developed. Now rather than considering aggregate demand for a geographic zone, the level-of-service variables for individual travelers and freight decisionmakers are used to make predictions of their choices -- so-called discrete choice models. These disaggregate predictions, of course, must ultimately be aggregated for the purposes of system design, but the idea that *behavioral* concepts can be used to generate more accurate demand projections is certainly progress.

Of course, we have not reached the ultimate in our ability to predict transportation demand. Indeed, many would argue it remains the most difficult modeling problem we face. And, while predictions are difficult from a technical point of view, the politically-based systematic overprediction of demands beyond what can be realistically supported in demographic or economic data remains a serious problem.

An example of how subtle demand predictions can be is the idea that travel time and, in particular, driving time in automobiles, is a dead-weight loss. People will naturally prefer shorter to longer trips; this concept drives the demand models that we use to make decisions about transportation infrastructure and operations. However, this factor is greatly leavened by the advent of telematics which permit driving time to be much more effectively used for business and social purposes. The cell phone is one example, and increasingly people are using their driving time for sophisticated business and social interactions through e-mail, fax and voice recognition systems for various purposes (to say nothing of telematics contributing to the speed and reliability of the trip itself through route guidance). We need to recognize that the fundamental quality of the driving trip has changed, which may importantly affect transportation demand predictions.

And, of course, as noted in (4), travel demand may be profoundly affected by substitutes available through telecommunications technology.

8.   **From**   **To**

Episodic Data for   **Dynamic Data for**
Investment Planning   ———————▶   **Investment Planning**
                                               (and Operations)

Traditionally, data for transportation planning purposes has been generated by an expensive, episodic data collection effort. That data is then used for planning for many years (because of the expense of the data collection activity itself), with that data becoming stale in the process. Now, with modern technology, we can collect data in a continuous manner at much lower cost. This permits us to use this data not only for operations management in real time, but also allows us to do investment planning on up-to-date data. An important value of new ITS technologies is in the ability of those systems to provide data and information for long-term strategic planning of the transportation network.

9.   **From**   **To**

                                            Private and Public/Private
Public Financing   **Partnerships for**
for Infrastructure   ———————▶   **Financing of Infrastructure**
and Operations   **and Operations**
                                            Using Hybrid Return on
                                            Investment Measures

Financing of both conventional and advanced transportation systems is forcing us to rethink institutional relationships. The role of the private sector in helping to fund transportation projects, long used in the developing world, is becoming important in the developed world as well. We see new kinds of public-private partnerships for the provision of transportation infrastructure and operations (e.g., concessions). Technology is central to these partnerships in that user fee recovery by the private sector is greatly eased through electronic toll collection.

Such hybrid projects need to combine the interests of both the public- and private-sector partners. We anticipate a transition from the use of different criteria for financing of infrastructure between the public and private sectors to the development of more *generalized return on investment* formulations which properly weight the interests of the private partners in making a profit and the public partners in providing for the "commonwealth".

10.

From	To
Infrastructure Construction and Maintenance Providers $\longrightarrow$	**New High-Technology Players**

The appearance of new (often high-tech) private-sector players in the transportation field is already prevalent. Communications, information technology and other high-technology organizations are now "transportation" providers. These companies often have different cultures than traditional infrastructure providers like construction companies and certainly have a better appreciation of the benefits of R & D.

Also, we are seeing new kinds of private business models. For example, cellular telephone companies work together with information service providers to supply information for travelers so they can effectively negotiate through a congested transportation network.

11.

From	To
Static Organizations and Institutional Relationships $\longrightarrow$	**Dynamic Organizations and Institutional Relationships**

Organizations, by their very nature, change slowly. The "first law" of organizations is self-preservation. It is not surprising, therefore, that many organizations in both the public and private sectors tend to change glacially, if at all.

In some of the transitions discussed above, we have already noted the need for change in organizations (e.g., from emphasis on capital facilities to operations, in developing a customer orientation, in sharing infrastructure finance responsibilities between the public and private sectors, and in incorporating new high-technology players and new business models). New partnerships between various governmental levels are important as well, as witness the interaction between federal and state organizations in the development of "regional architectures" for the deployment of ITS services.[7]

Further, consider the growing role of metropolitan planning organizations (MPO) (although these organizations vary substantially in power and effectiveness) as regional decisions come to the fore, as discussed in (3) above.

This suggests transportation organizations must change internally as well as in their institutional interactions with other organizations. *The people we*

---

[7] Pendleton, Todd and Joseph M. Sussman, "Regional Architectures: Strengthening the Transportation Planning Process", *Transportation Research Record 1679*, National Academy Press, Washington, DC, 1999.

educate to work in those organizations must be sensitized to the need for organizational change and new institutional relationships.

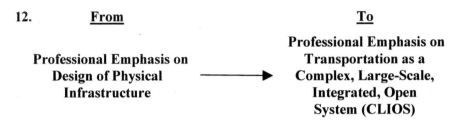

12.	**From**	**To**
	Professional Emphasis on Design of Physical Infrastructure	**Professional Emphasis on Transportation as a Complex, Large-Scale, Integrated, Open System (CLIOS)**

For decades the transportation profession was dominated by design of physical infrastructure. Issues such as pavement design, material selection, geometric design, earthquake-resistant design for highway bridges, etc., dominated the professional mindset. However, we have also long recognized, beginning nominally in the 1960s, that the professional practice of transportation needs to have a broader base in understanding the social, political, economic and environmental impacts of our decisions. Of course it continues to be important to design cost-effective, safe physical infrastructure, but *the broader perspective is now the sine qua non of transportation practice*.

In previous writings, we have introduced the idea of transportation as prototypical of complex, large-scale, integrated, open systems (CLIOS) as a way of capturing these ideas.[8] This broader perspective is essential in the analysis and design of modern transportation systems.

This new perspective on transportation is intimately linked with advances in modern technology, in particular, information technology. New computational capabilities enable us to model transportation systems as CLIOS. For example, we have the capability to solve enormous network problems, orders of magnitude larger than those we could deal with even a decade ago.

Of course, information technology is not the only enabler of considering transportation systems as CLIOS; we will require a new set of models and frameworks to consider the broad scope and scale of CLIOS, which recognize social, political and economic factors. Following from many of the above transitions is the notion that, as transportation professionals, we must recognize that the traditional civil engineering knowledge base, dealing with physical infrastructure, is not adequate for the future. Rather, we must be concerned with advanced technologies (information technology, telecommunications, sensors) and a broader CLIOS perspective on our field.

---

[8] Sussman, Joseph M., "The New Transportation Faculty: The Evolution to Engineering Systems", *Transportation Quarterly*, Eno Transportation Foundation, Washington, DC, Summer 1999. N.B. This article appears in Section III of this volume.

Finally, we note that not only is the physical system with which we are concerned a CLIOS. The process by which we make policy about it -- in investment, in terms of equity, in simply deciding how the system should be operated -- is itself a highly-complex institutional system. So we deal here with *nested complexity*, using a complex process to analyze, discern and design a complex system integrating institutional design and policy design with system design. The task is daunting.

13.   <u>From</u>   <u>To</u>

Economic   **Sustainable**
Development   **Development**

The transportation system has long (and appropriately) been thought of as a mechanism for economic development on urban, regional and national scales. Now we must broaden this concept to think of transportation as a mechanism for *sustainable* development, which includes recognition of the long-term effects of transportation on resources -- environmental, human, land, and so forth -- as well as economic development goals.

This has already manifested itself in the linkage of the Clean Air Act, with transportation legislation mandating that transportation system improvements must recognize the environmental impacts associated with those improvements. And technology permits a direct transportation/environment linkage through "environmental tolls", as discussed in (5) above.

Sustainability is simply a logical extension of concern with environmental impacts to a broader recognition that transportation is an important component of making urban areas, regions and nations sustainable, including such concepts as equity, wealth distribution, intergenerational perspectives, energy use, and so forth.

14.   <u>From</u>   <u>To</u>

Computers Are   **Ubiquitous**
"Just a Tool"   **Computing**

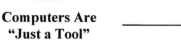

In the early days of the use of computers in engineering practice, dating back to the 1960s, we were at pains to explain that "computers were just a tool. While remarkably fast from a computational point of view, computers could never replace human decisionmaking." In a narrow sense, this is true. But back in the 1960s, and certainly today, it fundamentally understates the impact of computation on the world in which we live, and certainly on transportation systems. Fast and omnipresent computing (some call it *ubiquitous* computing), combined with very sophisticated and tiny sensors and telecommunications capability, has fundamentally changed the way we think about transportation systems and their design and operation.

"Just a tool"? Perhaps, but one that fundamentally changes our ability to understand, manage and control complex systems (transportation systems among others).

15.

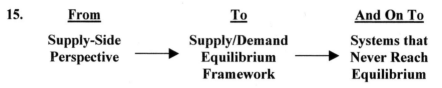

From	To	And On To
Supply-Side Perspective	**Supply/Demand Equilibrium Framework**	**Systems that Never Reach Equilibrium**

Our perspective on transportation has evolved from a simple supply-side perspective -- focused on providing transportation capacity through infra-structure development -- to, beginning in the 1950s, a more sophisticated notion of supply/demand equilibrium. This latter perspective recognized that demand for transportation services was a function of the level-of-service provided and that the level-of-service provided by a facility or network was itself a function of the actual volumes carried. The volume/capacity ratio is a determinant of level-of-service; as this ratio approaches unity, the level-of-service deteriorates substantially -- the so-called "hockey stick" pheno-menon.[9]

This notion of supply and demand as intertwined led to important work on *equilibrium* as a way of estimating and predicting transportation flows of both people and freight on complex networks. Indeed, the venerable four-step planning process (trip generation, trip distribution, mode split and path assignment) is undergirded by the equilibrium concept.[10]

More recently, we have begun to recognize that some systems *never* reach equilibrium. When we control an urban transportation network in real-time, while the system approaches equilibrium, in practice it never reaches that equilibrium state. Thinking in longer timeframes, some have argued that the freight transportation system of the U.S., governed by very long time constants related to the relative capacities and prices of the modal networks of rail, truck and inland waterways, fails to reach equilibrium even over a period of decades. This has given us pause about the predictions we develop using an equilibrium framework and has led to more sophisticated non-equilibrium methods for calculating flows, mode splits, and the like.

---

[9] Sussman, Joseph M., *Introduction to Transportation Systems*, Boston and London: Artech House Publishers, 2000.

[10] Sussman, Joseph M., *Introduction to Transportation Systems*, Boston and London: Artech House Publishers, 2000.

16.   **From**   **To**

Transportation firms and agencies have focused on conventional infrastructure, considered in a one-at-a-time, relatively independent manner. We recognize now that, particularly with advanced technologies as part of our infrastructure, we need to explicitly consider the linkages among these advanced infrastructure projects and the recognition that a *system architecture* is the basis for the overall design process.

System architecture is not a traditional concept in the transportation field. It draws from the ideas of computer-system architectures and complex defense weapons development. But, with the advent of ITS, we have recognized that such complex systems must be put together in an organized fashion using the concept of a system architecture which specifies:

1. the interconnections among components of complex systems from a communications and information processing perspective; and
2. where within a networked system we provide the information processing nodes and what information is communicated among nodes.

This is a basic part of the design of modern transportation systems and one that will enable the development of internally-consistent, interoperable systems.

Some researchers have extended this idea of an architecture so that it considers the interrelationship among the many transportation *organizations* operating at, say, the regional scale, to effectively work together. So the design of new *institutional networks* is also a part of this architectural concept.[11]

17.   **From**   **To**

Vehicles and   **Vehicles and**
Infrastructure as ⟶ **Infrastructure as**
Independent   **Electronically Linked**

For generations our highway design framework has treated the vehicle and infrastructure as independent subsystems. Certainly we had to be concerned with "where the rubber meets the road", but for the most part design of vehicles and design of infrastructure could proceed quite

[11] Sussman, Joseph M. and Christopher Conklin, "Regional Strategies for the Sustainable Intermodal Transportation Enterprise (ReS/SITE): Five Years of Research", Transportation Research Board, Washington, DC, 2001.

independently. With the advent of new technologies under the rubric of ITS, the vehicles and infrastructure are electronically linked. This "changes everything", particularly in the management and control of the transportation system. As noted earlier in (2), we now have the ability to monitor flows on the transportation network in real time and further to modify network operations to improve performance, as well as to provide traveler information to individuals to allow them to navigate the system more effectively.

Of course, as with many technological changes, there are institutional implications. Previously, traditional public-sector infrastructure providers and private-sector vehicle providers could proceed independently in their work. Now we require close coordination and cooperation between these organizations, to assure that the electronic linkages are handled properly, and that systems can be interoperable across broad geographic areas: nations or even continents. Vehicles and infrastructure being linked is of profound importance to the field of transportation on the dimensions of technology, systems and institutions.

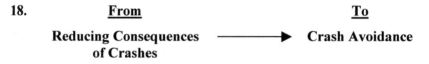

18.   <u>**From**</u>   <u>**To**</u>

Reducing Consequences ⟶ **Crash Avoidance**
of Crashes

Public- and private-sector transportation operators are often quoted as saying that "safety is our first priority". And, indeed, safety is fundamental. But our perspectives on safety are changing.

First, the highway safety community has long focused on ameliorating the effects of crashes when they occur. Seatbelts and airbags are technologies that are intended to minimize injuries and deaths resulting from crashes. In recent years technologies have permitted us to look at safety in a new way and focus more attention on *crash avoidance* through, for example, collision-warning-and-avoidance systems, intelligent cruise control, and like technologies.

Further, advanced technology can lower the probability of crashes through use of variable message signs warning of stopped traffic ahead, routing vehicles away from accident scenes where secondary accidents often occur. Finally, technology-based enforcement can remove dangerous drivers from highways (although privacy advocates have concerns here).

Second, in recent years we have seen a transition from rather ad hoc methods for deciding on appropriate levels of safety, to more quantitatively-based probabilistic risk assessment methods. These methods combine 1) the "impact" of an accident in terms of lives lost, personal injury and property damage, and 2) the probability of the accident occurring, into a coherent quantitative framework.

This perspective is not without controversy. Reflecting the value of human life and suffering in economic terms is anathema to many. But transportation professionals are increasingly recognizing the fact that, even if one does not *explicitly* specify these economic values in safety calculations for particular systems, there *is* a value they are *implicitly* assuming, whether they like it or not. Making these issues explicit can only help in the rational design of transportation systems.

19.

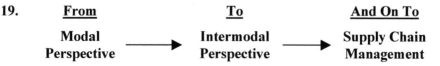

From	To	And On To
Modal Perspective	Intermodal Perspective	Supply Chain Management

A modal focus has been standard where people specialize, for example, in public transportation, highway transportation, air transportation, etc. And certainly we will continue to need such specialists.

But transportation is increasingly intermodal. In freight transportation the movements of containers internationally from truck to rail perhaps to truck again and ultimately to a container ship for international movement is the fastest-growing freight sector and a major driver of the global economy. The same is increasingly true in passenger transportation, where linkages among modes, taking advantage of the characteristics of each to create high-quality, low-cost services, are becoming increasingly important. Thinking explicitly about public transportation as part of an intermodal trip chain is an important example.

The Achilles' heel of intermodalism is traditionally the hand-off from one mode to another; the advantages gained in using the strengths of several modes are often dissipated in inefficient hand-offs. With new information, communication, and automation technologies, our abilities to handle hand-offs is much enhanced, allowing us to think more intermodally about transportation operations and services to meet customer needs.

Going beyond intermodalism, the concept of *supply chain management* explicitly recognizes the relationship of freight transportation to broader questions. Supply chain management considers the overall process of extraction of raw materials, to manufacturing of components, to an assembly into finished products, and ultimately transportation to the ultimate consumer. Some have defined it to include the appropriate disposal of products as well (e.g., automobiles). Clearly transportation is a critical element in this overall supply chain in virtually all its phases. This more sophisticated recognition of the role that transportation plays within this broader context represents an important evolution in the transportation field.

20.   From   To

Narrow Transportation   ➡   **The New Transportation**
      Specialists   **Professional**

As one looks at the transitions we have described above, it seems that the case is all the stronger for this *new transportation professional*, more broadly educated than the narrow transportation specialists of the past. We need people who

1. think broadly about Complex, Large-Scale, Integrated, Open Systems (CLIOS) and the integration of transportation system design with policy design;
2. can lead and work within organizations that are undergoing fundamental cultural change, as well as reaching out to other organizations in new and different ways;
3. have a strong customer orientation, recognizing that it is our customers who are our raison d'être;
4. understand the implications of new technologies on the design and operations of our transportation systems (as, for example, in understanding the long-term implications of ubiquitous computing in transportation system design);
5. understand how to effectively use the wealth of data now available to us for long-term planning, as well as for real-time management;
6. recognize that it is at the regional scale that the battles of economic competitiveness and environmental improvements will be joined and that transportation needs to be managed at that scale as well;
7. recognize that intermodalism (including communications as a "mode") is a mechanism for cost-effective and high-quality customer service, both in freight and traveler transportation; and
8. recognize our goal is a transportation system that contributes to a sustainable society, focusing on more than short-term economic and mobility goals.

CONCLUSION

So our world is changing. The transitions are many. The challenges are great. The education of the new transportation professional, as well as the re-education of the current one, is fundamental to our success in the future. While our field is "mature", it became so, and remains vital, because it has shown itself to be of fundamental importance socially, politically and economically. For transportation to continue to be effective in contributing on those dimensions, the new transportation professional -- the transportation professional of the 21st century -- must be able to deal with the many transitions described above (and doubtless some we have not yet anticipated).

V. 3. ITS: WHAT WE KNOW NOW THAT WE WISH WE KNEW THEN: A RETROSPECTIVE ON THE ITS 1992 STRATEGIC PLAN[1]

1.   ABSTRACT

From September 1991 until June 1992, a core writing team, which included this author, worked on what was the first Intelligent Transportation Systems (ITS) strategic plan in the United States. This plan was entitled, "A Strategic Plan for IVHS in the United States". It served to define the ITS program at a national scale in a way that has been characterized as seminal.

The plan, by most accounts, served as the blueprint for the early development of ITS in the U.S. and as the basis for the subsequent plans produced by ITS America, the federal government, various states, and a number of private-sector organizations.

This article explores various aspects of ITS retrospectively, contrasting views from 11 years ago, when the Strategic Plan was produced, with the current reality. Areas discussed include Advanced Traveler Information Systems (ATIS), Advanced Transportation Management Systems (ATMS), reliability, getting the ITS program off the ground in the early 90s, strategic use of information, automated network management, electronic toll collection (ETC), congestion pricing, architecture, commercial vehicle operations (CVO), Advanced Public Transportation Systems (APTS), and regions.

The article closes by comparing ITS with the Interstate system, and finally by discussing the upcoming reauthorization of the Transportation Efficiency Act for the 21st Century (TEA-21) and what has been learned through this retrospective about ITS-related issues on that reauthorization.

---

[1]   Reprinted with permission of the Transportation Research Board. Sussman, Joseph M., presented at TRB 83rd Annual Meeting, Washington, DC, January 2004, and accepted for publication in Transportation Research Record: Journal of the Transportation Research Board, 2004 (in preparation).

2.   PREFACE

From September 1991 until June 1992, the author had the unique privilege of serving on the core writing team for the ITS America (then IVHS America) Strategic Plan, entitled "A Strategic Plan for IVHS in the United States" (*1*). My sabbatical year from MIT gave me an opportunity to be at the very center of the development of the first Intelligent Transportation Systems (ITS) strategic plan in the U.S. It was a once-in-a-career opportunity to be in on a seminal moment of an emerging technology of substantial interest in my field of study. The writing team, with their respective affiliations at the time, was composed of Michael Sheldrick, the director of automotive business development for Etak, Inc., Menlo Park, California; Jonathan Arlook, the director of software engineering for the Navigation Technologies Corporation, Chicago; Edward Greene, the engineering design supervisor for the Electronics Division of the Ford Motor Company, Dearborn, Michigan; Craig Roberts, the director of institutional and legal issues for IVHS America (the precursor of ITS America); and me. We were charged with writing a good deal of the plan as well as coordinating inputs from a variety of committees and groups around the United States, from both the public and private sectors. The plan was subject to intense senior review (Thomas Deen, then TRB Executive Director; James Costantino, then President of ITS America; Lyle Saxton, FHWA; and William Spreitzer, General Motors), and ultimately was approved by the Coordinating Council and the Board of Directors of ITS America.

The plan, by most accounts, served as the blueprint for the early development of ITS in the U.S. and as the basis for the subsequent plans produced by ITS America, the federal government, various states, and a number of private-sector organizations.

Having said all this, it is clear that there was much the writing team and the ITS community more broadly defined did *not* understand about ITS and what would ultimately be the factors driving its success, or lack thereof. As I look back on that 1991/92 effort, while we did get much of it right, there was a good deal we got wrong as well. Some of these omissions were subtle; others were of the form "How could we (the nascent ITS community) not have thought of that?"

Of course, some of these omissions derive from the changes that have taken place in the world over the last 11 years that were to some extent unforeseeable in 1992. For example, while the Internet was beginning to have an impact, few foresaw the extraordinary change in workstyles and lifestyles inherent in that as an information/communication mechanism. The evolution of the environmental movement to one concerned more broadly with sustainability, including economic development, environmental impact, social capital, and so forth, was not yet understood. We knew that globalization was of fundamental importance, but the continued substantial growth of the global economy, fueled in large part by innovations in transportation systems, information systems and communication systems, was only partially foreseen. The focus that we have now on security in the post-9/11 era, did not exist. Demographers were pointing to the importance of the aging society back then. It is more recently that we have begun to really understand the implications of the aging of the baby boomers and their desire to retain lifestyle and mobility at previously experienced levels. And, the notion of developing-country megacities -- cities in excess of 10 million people -- was looming on the horizon, where now in 2003, it is established as one of the critical contemporary issues the world faces, from economic development, environmental, and equity points of view. Transportation is a key aspect of dealing with all these issues.

So the world has changed substantially in these 11 years and now, with the benefit of 20/20 hindsight, we can look back at what we did not fully understand about ITS and its implications, and what I hope we understand better now.

As I outline these issues, I should make clear that while I often use "we" or "the planners", these are really points that I personally did not grasp in the early days of ITS. While I am sure I was not alone in this lack of understanding, I am equally sure there are some who *did* understand back then. So, the reader should consider the points that follow as ideas the author "didn't get right"; I am not finding fault with the ITS community.

3.   ITS IN 2003: HOW IT COMPARES WITH THE 1992 STRATEGIC PLAN PERSPECTIVE

3.1   Advanced Traveler Information Systems (ATIS)

One of the subsystems of ITS that we thought we had a good understanding of was Advanced Traveler Information Systems (ATIS). We saw this as straightforward, with information being provided to drivers and transit users so they could make good decisions about how to access and utilize the transportation network.

What we did not foresee was the explosion in the methods for delivering traveler information to these drivers and transit users. The idea of the Internet as a source for traveler information had not occurred to us -- indeed, as Harold Worrall of Orlando-Orange County Expressway Authority has pointed out, the word "Internet" does not even appear in the 1992 Strategic Plan. The notion of ubiquitous use of cellular phones as a mechanism for receiving real-time traveler information during a trip was not fully appreciated. Indeed, these cell phones are now a mechanism whereby vehicles can be located and tagged on the infrastructure (e.g., through GPS) so travel times on segments of networks can be measured. This gives us a much better sense of the network state in the future, so important to giving good traveler information.

3.2 Advanced Transportation Management Systems (ATMS)

Advanced Transportation Management Systems (ATMS) is the companion subsystem to ATIS. While ATIS is directed to individual customers, ATMS is the subsystem through which we control or manage the transportation network for the benefit of the general public.

Our early view of ATIS and ATMS were as countervailing subsystems. ATIS was directed towards "relatively affluent" travelers who could afford to pay for special traveler information; ATMS was directed towards creating a better-operating transportation network for all drivers. The techniques that would be used included ramp-metering rates, dynamic variable message signs, incident detection and management, variable speed limits, and traffic light setting.

However, it now appears that perhaps the best mechanism that we have to improve overall transportation network performance is providing *better traveler information*, even if it is to a relatively small percentage of overall drivers. The other network methods mentioned above -- ramp-metering rates, etc. -- may have a positive impact on network operations, but the sense is that properly informed drivers, knowledgeable now about where real-time chokepoints in the network may be, and acting in their own self-interest to avoid them, may have the most substantial impact on improving network operations for *everybody* than any of the techniques described above.

So the notion of ATMS and ATIS as antithetical is wrong. They will likely rise and fall together rather than there being a trade-off between effective ATMS and ATIS. And since ATMS are usually public-sector operated and ATIS private-sector operated, this realization would seem to be quite important for developing public-private partnerships.

3.3   ATIS as a For-Profit Venture

The Strategic Plan envisioned private-sector organizations providing traveler information to individual users as a profit-making activity, absolutely necessary in the private sector. But at this writing, it is clear that making a profit in this business is very hard. As Jane Lappin of the Volpe National Transportation Systems Center (VNTSC) has said, "There is *no* market for traveler information."(*2*) (*3*) Free competition from radio stations makes it a very difficult sell; apparently the improvement in information that one gets from more sophisticated ATIS with a wider variety of data sources and a more real-time and route-specific nature is not something that many people are willing to pay for, at least not yet.

So while organizations are making money in ITS, it is mainly through the provision of ITS infrastructure to the public sector rather than through the sale of better traveler information to the public in general.

3.4   Reliability

In the strategic planning exercise of 1991/1992, we emphasized the benefits in improved travel times to drivers as a key economic benefit of ITS.   Time is money, we said.   Getting there faster has economic value. While this is true, we overlooked, until quite recently, the importance of reliability to the highway traveler. Reliability is a measure of the variability in travel time between two points. We are all familiar with the fact that on Monday we have a quick half-hour trip from origin to destination, but on Tuesday an accident or storm or construction may cause that same trip to take twice that time. If one is risk-averse about being late, one must build additional time above the "quick" travel time into one's time budget. Often it is wasted time in the sense that one arrives at the destination earlier than one intended. But it is a price one may be willing to pay to avoid, for example, lateness for an important meeting (or even a class!).

In many cases, improved reliability available through real-time information about *today's* trip time is proving to be more important than improvements in average travel time. This has been a phenomenon that has been well understood for decades in freight transportation. (*4*) The trucking industry has won considerable traffic from the rail industry, even charging premium rates, because they provide more reliable trip times.   This is because unreliability generates additional inventory costs for the customer. However, the understanding of the importance of reliability for highway travelers, where time management is critical, is relatively recent. (*5*)

Indeed, it now appears that actual highway travel time savings are often ephemeral or rather small. There is little empirical evidence to show that the small improvements in average travel time are economically meaningful. When people get more reliable trips by receiving information about expected travel time in real time before the trip, some suggest they end up unconsciously converting that into (often non-existent) travel time savings in their minds. What they have actually accomplished has been better *time management* when they receive real-time information about a trip that is about to be on the right-hand tail of the travel time distribution (or even the left-hand tail!).

3.5   Getting the ITS Program off the Ground

In the 1991/92 strategic planning era, there was a lot of discussion of "Alphonse and Gaston". The planners were concerned that the private sector would not make R&D investments in in-vehicle equipment until they were certain that the public sector would roll out public ITS infrastructure. Conversely, there was concern that the public sector would not do that roll-out unless the private sector had a commitment to in-vehicle equipment.

This concern was ill-founded. In the U.S., with a very strong Federal Highway Administration in the lead and holding most of the cards (and dollars), the public sector clearly took the initiative in rolling out ITS; the private sector, especially the automobile manufacturers in the United States, was more lethargic in developing in-vehicle equipment, and probably lags to this very day behind Japanese and European carmakers.

It is interesting that this same pattern of roll-out did not occur in Japan. There, rather than an omnipotent Federal Highway Administration, there was a war of the ministries -- including the Ministry of Transportation, the Ministry of Post and Telecommunications, the National Police Agency, and so forth -- about dominance in their nascent ITS movement. Eventually, the private-sector automobile manufacturers tired of this bureaucratic arm-wrestling and rolled out their autonomous in-vehicle systems, without waiting for a public-sector commitment to ITS. This demonstrates how institutional form can create differences in the way technologies are developed, as documented by Hans Klein in his MIT doctoral thesis. (6)

3.6   Strategic Use of Information

In 1991/92, the emphasis in the "IVHS" Strategic Plan was the collection of data about traffic conditions in real-time that could lead to more efficient network flows and improving, through traveler information, trips for individual travelers. Of course, this has happened. But what was largely overlooked was the *strategic use* of this same information for transportation planning purposes.

As the author has noted in a previous article (7), prior to ITS, data for strategic network planning -- the adding of infrastructure most particularly -- was ofttimes based on relatively ancient information because the cost of collecting data was so high. Now, with intelligent infrastructure in place collecting information for real-time operations, we have as a quite important side benefit the development of archival data that can greatly improve the quality of planning for strategic network change. Like other notions in this article, this is clear in retrospect. At the time, our focus was so much on the "modern" applications of real-time data that we largely overlooked the more conventional advantages of large-scale transportation databases that came "for free" with the development of ITS infrastructure.

3.7   Automated Network Management

The gleam in the eye of the strategic planners in the early 90s was an automated system that collected data in real-time from transportation infrastructure and vehicles, and then, through *intelligent algorithms*, made automatic changes in network operations (i.e., without human intervention), so as to improve traffic flows and provide a better (i.e., faster) trip for travelers. The gathering of real-time data has happened, of course, but algorithms that change network operations have been slow to develop; to the author's knowledge, it is virtually non-existent. Indeed, the author argues that the only ubiquitous automated transportation network management is through traffic light systems that are automatically modified in real-time without human intervention (SCOOT, SCAT) (8). Other than that, the author believes that all the other network systems are *decision-support systems* for human decisionmakers who look at that data, presumably clearly presented for them, and then make a (human) decision about what to do to enhance network operations. The next evolutionary step, when the computer makes that choice, essentially does not yet exist.

We suspect that someday this will be the case. But, if someone had told us in 1992 that in 2003 this would still be a will-o'-the-wisp, we would have been dismayed.

3.8   Electronic Toll Collection (ETC)

Another example of the slowness of development that would *not* have been predicted by the strategic planners in the early 1990s is electronic toll collection (ETC). You may be surprised to hear this; certainly many (including me) would argue that electronic toll collection is the major success story of ITS, with implementations all over the country and abroad. But there are still many states without any electronic toll collection, when the technology has long since been proven. *Who thought it would take so long?* Certainly not the strategic planners.

What is also disquieting is the lack of compatible electronic toll collection systems, even in regions with many small states like New England, where it makes overwhelming sense. The inability of organizations in the public sector to cooperate in the development of common technologies for the convenience of the traveling public continues to be a major barrier to compatible electronic toll collection systems. It is relatively recently that the E-Z Pass system in the New York metropolitan area was implemented after a good deal of "negotiation" among the states of Connecticut, New York and New Jersey. And it is even more recent that E-Z Pass has been made compatible with the FastLane system in Massachusetts. E-Z Pass or FastLane only now is becoming available in New England states other than Massachusetts or Connecticut, where New England, with a number of small states, would really gain from compatible deployment of ETC.

Most ambitiously, we would certainly hope that a nationally-scaled compatible ETC system would be in place. Truckers, for whom long trips across political boundaries are common, would doubtless find this of great value. One could imagine a single transponder in rental cars, where the tolls could simply be added to the bill rather than the driver fumbling for change. But the current reading on getting a truly national system, which the strategic planners in the early 90s viewed as important (and even straightforward), is that it is a long way off due to that old bugaboo: *institutional issues*.

The strategic planners were not naïve about institutional issues; we realized they were going to be very difficult in the deployment of a new technology in a conservative industry. But it is fair to say we grossly underestimated just how difficult it would be. An anecdote: In a talk at the ITS Massachusetts Annual Meeting in 2003, we heard about two variable message signs in rural central Massachusetts obtained "for free" through federal funding not being deployed for more than a year because of bureaucratic quibbling between two small public organizations about who would "really" own and operate them.

3.9   Congestion Pricing

Congestion pricing, or value pricing as it has also come to be called, was viewed as an important application of ITS technology at the time of the 1992 Strategic Plan. It had the potential to smooth the peaks on congested highways by allowing individual drivers to make a choice about whether they were willing to pay a premium for traveling at a particular hour. It was an idea that Professor William Vickrey first put forth in the 1950s; it was part of the body of work that won him the Nobel Prize in Economics in 1996. Finally it had become technically feasible, based on ETC technology.

Many of us, going back even to the early 90s, have been saying that "congestion pricing is inevitable", but in fact it has taken a long, long time. Earlier this year (2003), London instituted congestion charging in central

London, with virtually all drivers paying about $8/day to cross the cordon. Early reports are that this has had a substantial positive effect on congestion. The notion is that if London can do it, perhaps many other cities can also (an argument that does not work for Singapore's success, given their special political environment). So maybe the dam on congestion pricing will break now, but again the timeframe has been much longer than the planners thought.

In the U.S., the idea of high-occupancy toll (HOT) lanes, where single-occupancy vehicles can use the HOT lanes if they are willing to pay a toll, is an important application of value pricing. We expect to see more such applications in the future. But, frankly, we expected to see these in the mid-1990s!

Another surprise to the planners is who is using the HOT lanes. The conventional wisdom is that the wealthy would "unfairly" take advantage of this service. In fact, working women (not necessarily wealthy) are disproportionate users as they (in our current culture) try to manage professional and personal responsibilities (so facing confiscatory penalties for being late to the day-care center, a HOT toll seems a small price to pay.) (9)

3.10 Architecture

Developing a system architecture for ITS was recognized in the 1992 Strategic Plan, but I do not think any of the planners recognized the extraordinary effort this would become, *and* certainly would not have predicted the ultimate use of the architecture.

The ITS architecture was developed in a "fly-off", well-known to DOD and weapons-system development, but virtually unknown to the U.S. DOT. In 1994, four companies, Hughes, Loral, Rockwell and Westinghouse, each with various subcontractors, developed competitive architectures and ultimately the best was selected as the basis for the ITS architecture. (10) U.S. DOT more normally would have selected the "best" contractor and used the architecture they developed. But this was a special situation. ITS was the first civilian surface transportation system viewed as technologically complex enough to require an architecture. Some were, and still are, skeptical that it does require that kind of top-down system design concept. In any case, the architecture became a part of the ITS world.

The notion of "regional architectures" that were required (by FHWA) to be "consistent" with the national architecture came later. I believe it is fair to say there is still not a complete understanding in the ITS community of what the term "consistent" means in this context. This author wrote a column in 1999 trying to clarify what "consistency" meant. (11) Others joined the battle, but there is still some confusion.

The Federal Highway Administration has used this consistency concept as a mechanism for controlling funds flowing from the federal government to the states. I think it is fair to say that regional architectures have become a negotiating ground for various public-sector transportation organizations to develop ITS in their regions. It is not clear that it has led to better *integration of transportation operations*, when they are run by a number of different public-sector agencies, as is usually the case.

So, while in the 1992 Strategic Plan architecture was noted as a necessity for ITS because of its technical complexity, the energy and resources that went into the development of this architecture and its current use, now focused more on administrative control rather than technical advance and integration, is a surprise.

Also, the *routinization* of the development of architectures through specialization software such as TurboArchitecture (*12*) may have had a negative impact on how much deep thinking goes into the development of an *integrated*, high-technology surface transportation system. Moreover, the use of TurboArchitecture and other shortcuts may delay the retooling of organizations and interorganizational relationships so necessary for effective deployment of an integrated ITS system.

In the author's view, the architecture grew from a straightforward concept to structure a technical system into a massive effort emphasizing administrative control.

On the other hand, some researchers, including the author, have redefined the regional architecture as an *organizational* concept, for specifying information flows and control hierarchies among participating organizations, with some good results. (*13*)

So, the term "architecture" has come a long way from the concept in the 1992 Strategic Plan, in some ways positive but in some ways less so.

3.11   Commercial Vehicle Operations (CVO)

The view from the strategic planners was that commercial vehicle operations (CVO) could be an early winner in ITS. Real-time routing of trucks, built on automatic vehicle location (AVL) technology for large truck fleets, was viewed as a mechanism for enhancing productivity. Since trucks are a private-sector enterprise, these productivity enhancements could come directly down to the bottom line. So if UPS, for example, could serve a metropolitan area with 40 rather than 50 trucks providing the same level of service, due to productivity enhancements, that 10 fewer trucks manifests itself in more profit, or perhaps lower costs for shippers. It is fair to say that these technologies have had a positive impact on commercial vehicles and their operation. The planners did not foresee the negative reaction in the trucking industry, which was quite concerned about the federal government having information about their operations for privacy reasons, and also because it might leak to competitors -- and many firms basically said to the ITS community, "We're already doing this; stay out of our hair." Many of these issues have been worked out by now, but it was difficult in the early years.

The other aspect, largely unheralded by the planners, was the substantial positive benefit of the *automation* of *mundane transactions* between truck companies and state regulatory agencies ("one-stop shopping"). Some thought this was not really ITS -- payment of excise taxes, relicensing, and so forth -- and perhaps it is not, but the automation of these transactions has had a quite positive benefit through the CVISN system throughout the United States. (*14*) Its importance was underestimated by the planners in the early 1990s.

3.12   Advanced Public Transportation Systems (APTS)

Some friction existed during the development of the Strategic Plan between the Federal Transit Administration (FTA) and the planners. The FTA felt that initial drafts of the plan were focused almost entirely on highway applications and did not give proper weight to the applications of ITS to public transportation. They were right. After all, at that time we were talking about "Intelligent Vehicle Highway Systems (IVHS)"! Eventually the Strategic Plan had a major section describing Advanced Public Transportation Systems (APTS) and APTS came to be thought of as a potential early winner of ITS. It could improve fleet productivity (as with the trucks noted above) and lower costs for transit properties, as well as provide a higher quality of service for transit travelers through headway control and through real-time traveler information.

Unfortunately, the rhetoric did not match the action in the field. APTS have been much slower to develop than the FTA hoped. (*15*) There are many reasons for this. Transit agencies tend to be cash-poor and risk-averse. Often they do not have the staff necessary to evaluate and then utilize high-technology systems. We all hope that someday APTS will have a substantial effect nationwide on public transportation, but currently the rhetoric outstrips the accomplishment.

3.13   Regions

Operating at a regional scale in surface transportation was an early dream of the strategic planners. (*16*) (*17*) Today it is even fair to say that there is a consensus -- in principle -- that ITS gives us the capability of operating effectively at a regional scale, a scale much geographically larger than was feasible in a pre-ITS era. The idea of the regional architecture described above is one manifestation of this consensus about regionalism and transportation. Applications such as Transcom in the New York Metropolitan Area and TransInfo in the Bay Area are proof that it can be done. But for the most part, regionalism has faltered under the difficulties in overcoming many of the institutional issues in cooperating on transportation needs 1) between the inner city and suburban communities; 2) between states in multi-state regions; and 3) between public safety and transportation operations organizations.

There is no question that a strong theoretical case can be made for regionally-scaled operations in terms of effectiveness and efficiency but, as with many other ITS concepts, the slowness to develop on the ground is a disappointment to the early advocates and planners.

3.14   ITS Compared with the Interstate

During the creation of the 1992 Strategic Plan, much rhetoric dealt with the equating of ITS with the Interstate System (e.g., ITS as a 21st century equivalent of the Interstate System). In terms of impact, some planners felt ITS and the Interstate would one day be comparable. It is, of course, early in the game, but thus far the Interstate has had much more profound impacts, both good and bad. The Interstate represented a fundamental change in mobility in the U.S. and helped create enormous economic growth. Nonetheless, it has its critics who speak of adverse effects in cities (e.g., destruction of neighborhoods and urban fabric, sprawl), environmental impacts, and equity considerations. What no one disagrees about is the *magnitude* of the effects of the Interstate on many dimensions.

ITS is, of course, much younger; perhaps the major effect of strategic interest is the use of ITS technologies for supply chain management and freight logistics (and some would argue that movement pre-dated ITS anyway). We have not seen a major shift of infrastructure expenditure in highways from conventional infrastructure to the high-tech infrastructure that ITS represents, but this may yet come. (I note that the current Bush administration has not bought into the early ITS rhetoric of "you can't build your way out of congestion" with a big conventional infrastructure program.) In-vehicle equipment (telematics) has had some effect on driving behavior, but not anywhere nearly as importantly as the Interstate did. Electronic toll collection has been a big-impact item, but it could be argued that this is only a convenience at the margin rather than a basic change. Congestion pricing, which *would be* a profound change, has yet to become prevalent, although the signs are beginning to be positive (London congestion charging scheme, as discussed earlier, for example).

So, 11 years after the 1992 Strategic Plan, I would argue the strategic impacts of ITS still lie before us and, certainly, we have a long way to go before its impacts approach those of the Interstate.

3.15   Security

As noted earlier, back in the early 90s, 9/11 was an unpredictable nightmare, and the role of ITS in security was, if stated at all, a modest add-on to a technology focused on safety enhancements and congestion improvements. But now in 2003, almost two years after 9/11, the use of ITS as a tool for enhancing national security is front-and-center on the agenda. Certainly these concerns will make ITS a more saleable technological concept. This is important as we approach the reauthorization of the "Transportation Efficiency Act for the 21st Century" (TEA-21). The allocation of funds for ITS in the reauthorization legislation will have an important impact on future deployment. And this leads me, finally, to a more general comment on that reauthorization and the implications for ITS.

4.   REAUTHORIZATION OF TEA-21

How can we assure the continued effective deployment of ITS technologies? TEA-21 reauthorization is an instrument that is quite important for the future success of the transportation enterprise in the U.S. And allocation of funds to ITS is an important factor for the future of that innovative segment of the transportation system.

From almost the beginning of ITS, the conventional wisdom has been that ITS must demonstrate *"real benefits"* before it can be fully accepted. I am not sure that is completely true. The history of transportation investment says otherwise. Here we are, essentially finished with the Interstate; yet there is still substantial debate about the benefits that accrue to society as a result of that extraordinary infrastructure deployment, and whether those benefits have outweighed costs that the environmental community would point to. Even the railroads, whose major years of building took place in the late 19th and early 20th centuries, have had their impact questioned and, to this day, the role of public finance of those major infrastructure projects is debated.

It is unlikely we will every be able to build an ironclad case for ITS benefits. Indeed, we may still not fully understand what the benefits are! For example, as noted earlier, a recent finding has been that it is reliability and not average travel time that matters in what ITS provides, where for the past decade we have considered improved travel time as the sine qua non of ITS benefits.

But I believe we can build political, professional and public acceptance of ITS. A better job of linking early successes like electronic toll collection to the more general uses of technology in surface transportation is a political approach. Politicians need help from us in characterizing ITS as a *project* to their constituents in the same way conventional infrastructure is viewed. The development of the "New Transportation Professional" (*18*), with more technologically-sophisticated people coming into the transportation industry, can lead to professional acceptance of ITS. Taking advantage of the aging baby boomer demographics, that generation's political power, and the desire of that generation to retain their mobility as long as possible is clearly a way toward public acceptance. By allowing people to continue to drive as they age, ITS can build support.

The strategic question for the ITS community is the extent to which bundling these political advantages and early winners into an integrated ITS program is the best political approach, as contrasted with selling each ITS component on its own. Certainly the early strategic planners saw an integrated approach as an advantage. We need now to take a hard look at whether it continues to be so.

5.   A FINAL WORD

So, looking back on the 1992 Strategic Plan, while it served as a base for what came later, there were obvious things we overlooked; there are things that really happened, although over a much longer time period than we had hoped; we underestimated some factors such as institutional issues (despite believing we were conservative in our time estimates); there were concepts such as the architecture that greatly expanded from the initial notion of the need for a technical architecture into a concept with a life of its own with more administrative than technical content.

But good progress has been made in ITS. The field has matured. The profession is changing. And in all we see ITS as an integral part of our surface transportation systems, building on our current successes for years to come. *But*, patience, as always, will continue to be required as we wait for the full impacts of ITS to be felt.

References

1. *A Strategic Plan for IVHS in the United States*. IVHS-AMER-92-3. IVHS America, Washington, DC, May 1992.
2. Lappin, J. E. What have we learned about advanced traveler information systems and customer satisfaction? Chapter 4 in *What Have We Learned About ITS?* Federal Highway Administration, U.S. Department of Transportation, Washington, DC, December 2000.
3. Lappin, J. Advanced Traveler Information Services (ATIS): Who Are ATIS Customers? Presented at the ATIS Data Collection Guidelines Workshop, Scottsdale, AZ, February 2000.
4. Sussman, J. M., A. S. Lang and C. D. Martland. Reliability in Railroad Operations: Executive Summary. Vol. 9, R73-4, Department of Civil Engineering, Massachusetts Institute of Technology, Cambridge, MA, December 1972.
5. Wunderlich, K., M. Hardy, J. Larkin, and V. Shah. On-Time Reliability Impacts of Advanced Traveler Information Services (ATIS): Washington, DC, Case Study. U.S. Department of Transportation, ITS Joint Program Office, Washington, DC, January 2001, EDL #13335.
6. Klein, H. Institutions, Innovation, and Information Infrastructure: The Social Construction of Intelligent Transportation Systems in the U.S., Europe, and Japan. Doctoral Thesis, Massachusetts Institute of Technology, Cambridge, MA, June 1996.
7. Sussman, J. M. Transitions in the World of Transportation. *Transportation Quarterly*, Vol. 56, No. 1, Winter 2002, Eno Transportation Foundation, Washington, DC, 2002.
8. *Traffic Control System Handbook*. U.S. Department of Transportation, Washington, DC, 1996.
9. Supernak, J., Golob J., Kawada K. and T. Golob. San Diego's I-15 Congestion Pricing Project: Some Preliminary Findings. Presented at the 78th Annual Meeting of the Transportation Research Board, Washington, DC, 1999.
10. Parsons, R. Issues in Developing and Implementing the National ITS Architecture. Chapter 19 in *Intelligent Transportation Primer*. Institute of Transportation Engineers, Washington, DC, 2000.
11. Sussman, J. M. Regional ITS Architecture Consistency: What Should It Mean? Thoughts on ITS Column. *ITS Quarterly*, ITS America, Washington, DC, Fall 1999.
12. Turbo Architecture Version 2.0, Federal Highway Administration (FHWA), McTransTM Center for Microcomputers in Transportation, University of Florida, May 2002.
13. Sussman, J. M., and C. Conklin. Regional Strategies for the Sustainable Intermodal Transportation Enterprise (ReS/SITE): Five Years of

Research. In *Journal of the Transportation Research Board, No. 1747*, TRB, National Research Council, Washington, DC, 2001.

14. Bapna, S., J. Zaveri, and Z. A. Farkas. *Benefit/Cost Assessment of the Commercial Vehicle Information Systems and Networks in Maryland*, National Transportation Center Morgan State University, Baltimore, MD, EDL No. 9369.

15. Casey, R. F. (Volpe National Transportation Systems Center). What have we learned about advanced public transportation systems? Chapter 5 in *What Have We Learned About ITS?* Federal Highway Administration, U.S. Department of Transportation, Washington, DC, December 2000.

16. Sussman, J. M. ITS Deployment and the 'Competitive Region'. Thoughts on ITS Column. *ITS Quarterly*, ITS America, Washington, DC, Spring 1996.

17. Sussman, J. M. *Transportation Operations: An Organizational and Institutional Perspective*. Report for National Special Steering Committee for Transportation Operations and Federal Highway Administration/U.S. Department of Transportation, http://www.ite.org/NationalSummit/index.htm., Washington, DC, December 2001.

18. Sussman, J. M. Educating the "New Transportation Professional". *ITS Quarterly*, ITS America, Washington, DC, Summer 1995.

Afterword

"Innovation is at the core of creating a sustainable human society. As a society, we will not succeed in creating a sustainable world if we focus merely on doing more efficiently what we currently do."
-- "Innovation, Technology, Sustainability & Society"
World Business Council for Sustainable Development

I hope you enjoyed this tour through ITS, ranging from descriptions of ITS and its subsystems, extending through perspectives on how ITS has affected transportation, where we are today and what the future portends. The transportation system is certainly a Complex, Large-Scale, Integrated, Open System (CLIOS) (Sussman, Joseph M., *Introduction to Transportation Systems*, Artech House Publishers, Boston and London, 2000) and ITS is a complex system within it. It has far-reaching implications on the dimensions of technology, systems and institutions for the world of transportation and beyond.

This book tries to capture the richness of ITS. My hope would be that it gives the reader insight into what ITS can contribute as a technology, its impact on many related systems and organizations, and how we can harness the ITS concept, broadly defined, to help create a sustainable society.

Index